# Information Security and Ethics:

## Social and Organizational Issues

Marian Quigley
Monash University, Australia

**IRM Press**

Publisher of innovative scholarly and professional
information technology titles in the cyberage

Hershey • London • Melbourne • Singapore

| | |
|---|---|
| Acquisitions Editor: | Mehdi Khosrow-Pour |
| Senior Managing Editor: | Jan Travers |
| Managing Editor: | Amanda Appicello |
| Development Editor: | Michele Rossi |
| Copy Editor: | Ingrid Widitz |
| Typesetter: | Amanda Appicello |
| Cover Design: | Debra Andree |
| Printed at: | Integrated Book Technology |

Published in the United States of America by
      IRM Press (an imprint of Idea Group Inc.)
      701 E. Chocolate Avenue, Suite 200
      Hershey PA 17033-1240
      Tel: 717-533-8845
      Fax: 717-533-8661
      E-mail: cust@idea-group.com
      Web site: http://www.irm-press.com

and in the United Kingdom by
      IRM Press (an imprint of Idea Group Inc.)
      3 Henrietta Street
      Covent Garden
      London WC2E 8LU
      Tel: 44 20 7240 0856
      Fax: 44 20 7379 3313
      Web site: http://www.eurospan.co.uk

Library of Congress Cataloging-in-Publication Data

Information security and ethics : social and organizational issues / Marian Quigley, editor.
     p. cm.
  Includes bibliographical references and index.
  ISBN 1-59140-233-6 (pbk.) -- ISBN 1-59140-286-7 (hardcover) -- ISBN 1-59140-234-4 (ebook)
 1. Technology--Social aspects. 2. Computer security. 3. Information technology--Social aspects.
4. Information technology--Moral and ethical aspects. I. Quigley, Marian.
  T14 5 I55 2005
  303.48'3--dc22
                                   2004003767

British Cataloguing in Publication Data
A Cataloguing in Publication record for this book is available from the British Library.

All work contributed to this book is new, previously-unpublished material. The views expressed in this book are those of the authors, but not necessarily of the publisher.

# Information Security and Ethics:
## Social and Organizational Issues

# Table of Contents

# Preface

*Information is invisible, communicative, and laden with value and ethical implications* (Baird et al., 2000, p. 10).

Drawing on the earlier writings of Joseph Weizenbaum (1976), Stacey L. Edgar, in the introduction to his excellent book, *Morality and Machines,* emphasises the need to "examine the dangers of being too mesmerised by the 'computational theory of mind,' which can, with its deterministic and materialistic implications, lead to losing sight of what is of moral (and aesthetic) *value*" (2003, p. 7). Similarly, the renowned communications theorist Raymond Williams reminds us that "a technology is always, in a full sense, social. It is necessarily in complex and variable connection with other social relations and institutions ... " (1981, p. 227). It is in a timely manner, therefore, that *Information Security and Ethics: Social and Organizational Issues* brings together a collection of recent work by international scholars addressing a number of significant and current social and moral issues associated with the development and use of new information and communication technologies.

The interrelated areas of information security and information ethics are rapidly gaining importance in the wake of the terrorist attacks on the USA on September 11, 2001 and at the same time as academics, computer professionals, government agencies, business organisations and the general public are becoming increasingly aware of the dangers associated with our growing reliance on computer technologies    particularly with regard to the ubiquitous and unregulated nature of the Internet. Today, all members of society are affected by computers - even if they themselves do not own one. The computer has changed our home and workplace environments, how we communicate, how we do business, how we shop and how our children are educated and entertained. As parents, we may be becoming more and more concerned about our inability to protect our children from what we perceive as the harmful effects of technology. As citizens, we may be growing increasingly anxious about the external threats to our national security posed by cyber-terrorists and the internal threat to us as individuals of government control and the related invasion of our rights to privacy and free speech. Business organisations, meanwhile, need to be constantly alert to the increasing dangers to their information security and intellectual property posed by hackers and white-collar criminals. The responsibility lies not just with managers. Kevin Day suggests that had all employees been educated in security then the majority of recent successful security attacks could have been avoided. "Security is not a technology; it is a thought process and a methodology" (Day, 2003, p. 4).

An uncritical acceptance of technological growth and development can and has provoked a series of ethical dilemmas. As James Moor warns us,

> *Computer sprawl, like urban sprawl, moves inexorably on many fronts unsupervised ...[It] is worldwide and culturally transforming. Computer sprawl is not necessarily rational or harmless, but it is an undeniable force in the world that will affect not only the lives of all of us in technological societies but quite possibly everyone on the planet and their descendants for centuries to come. The ethics gap that is generated because we massively computerize without taking time to consider the ethical ramifications is therefore quite wide and deep* (Moor in Baird et al., 2000, pp. 35-36).

While the computer-literate amongst us may enthusiastically embrace new technological developments and the associated changes they bring to our lives, other groups such as the less educated, the aged, the disabled and those living in less developed nations are becoming increasingly marginalised and powerless. Just as C.P. Snow in 1959 alerted us to the dangers inherent in the gap between the members of the "two cultures" of the sciences and humanities within Western societies, a number of the authors in this volume attest that the world-wide gulf between the information-rich and the information poor - the "two cultures of the computer age" (Weizenbaum in Edgar, 2003, p. 2) - is rapidly widening. In order to achieve global ethical solutions to this major problem, "[members] of the scientific community must bring a greater technical understanding of the underpinnings of the technologies involved; those from the humanities must bring a basis on which to make moral judgements and choose social and political alternatives well" (Edgar, 2003, p. 3). As Edgar suggests, we would do well to consult the writings of founding moral philosophers such as Aristotle in order to establish an ethical framework adaptable to the Information Age.

At the same time, we need to be aware that, because of the rapid development of technology, it seems likely that there will always be an ethics gap between technology and its use (Baird et al., 2000). Nonetheless, it is through the ongoing, "cross-cultural" (that is, between the sciences and humanities as well as between local/regional cultures) and inter-disciplinary studies, debates and discussions undertaken by international scholars such as those present in this volume, that, hopefully, we may achieve greater global awareness and possible solutions to the ethical dilemmas we are now facing within technologised societies.

# Organisation of the Book

The content of the book is organised into two parts: Part 1 (Chapters 1-8) focuses on Information Ethics; Part 2 (Chapters 9-14) focuses on Information Security. A brief description of each of the chapters follows:

## Part I: Information Ethics

Chapter I: *MAMA on the Web: Ethical Considerations for Our Networked World* by Barbara A. Schuldt, Southeastern Louisiana University, USA, addresses the need to find a global solution to the ethical problems caused by the open nature of the Internet and the growth and diversity of its users. To this end, the author proposes the adoption of four categories to enable the definition and discussion of these ethical issues. This chapter provides a framework for the discussion provided by the authors of the following chapters.

Chapter II: *Establishing the Human Dimension of the Digital Divide* by Helen Partridge, Queensland University of Technology, Australia, considers the psychological factors contributing to the digital divide through an examination of Internet users and non-users in Brisbane, Australia and San Jose, California, USA. Through this study, the author aims to expand the understanding of this phenomenon in order to enable the development of strategies and programs to bridge the gap between the information rich and information poor.

Chapter III: *Socio-economic Influence on Information Technology: The Case of California* by Rasool Azari and James Pick, University of Redlands, USA, proposes the steps that need to be taken by the State of California in order to foster technology and reduce the digital divide. The authors emphasise the inter-relationship between socio-economic factors and technological development, warning that focusing solely on the equality of distribution of technologies is insufficient in solving the digital divide.

Chapter IV: *The Ethics of Web Design: Ensuring Access for Everyone* by Jack S. Cook, Rochester Institute of Technology, USA, and Laura Cook, State University of New York, USA, addresses the problem of Web accessibility for the disabled — including the aged and injured. The authors emphasise the need to educate Web designers — many of whom are unaware of the problem, as well as the need for laws enforcing Web access for all.

Chapter V: *Web Accessibility For Users with Disabilities: A Multi-faceted Ethical Analysis* by Alfreda Dudley-Sponaugle and Jonathan Lazar, Towson University, USA, analyses the ethics of Web accessibility and argues that it is ethical to provide access for the disabled and unethical to exclude them. The authors suggest that the general population would also benefit from more accessible Web pages.

Chapter VI: *Internet Voting: Beyond Technology* by Trisha Woolley and Craig Fisher, Marist College, USA, discusses the issues of privacy, security, authentication and access associated with Internet voting and concludes that its benefits are currently outweighed by negative factors.

Chapter VII: *Protection of Minors from Harmful Internet Content* by Geoffrey A. Sandy, Victoria University, Melbourne, Australia, addresses the problem of how to protect minors from Internet material perceived as harmful without violating adults' right to free speech. The chapter provides a critique of Australia's regulatory framework.

Chapter VIII: *Mobile Communities and the "Generation that Beeps and Hums"* by Marian Quigley, Monash University, Berwick, Australia, argues that, at a time when critics are debating the demise of community, young people today are utilising mobile phones — alone or in combination with the Internet — to establish and maintain mobile, peer-based, social networks.

## Part II: Information Security

Chapter IX: *Insights from Y2K and 9/11 for Enhancing IT Security* by Laura Lally, Hofstra University, USA, analyses the Y2K and 9/11 disasters, showing how current Information Technology (IT) infrastructure allows for the propagation of IT threats. The chapter also analyses the efficacy of available IT tools in identifying potential security threats and in mitigating their impact.

Chapter X: *Cryptography: Deciphering Its Progress* by Leslie Leong and Andrzej T. Jarmoszko, Central Connecticut State University, USA, argues that the increase in cyber-terrorism, hackers and white collar crime highlights the need for a stronger security measure in cryptography.

Chapter XI: *A Method of Assessing Information System Security Controls* by Malcolm R. Pattinson, University of South Australia, Adelaide, Australia, proposes a method for assessing a small business organisation's Information System security, utilising an Australian case study.

Chapter XII: *Information Security Policies in Large Organisations: The Development of a Conceptual Framework to Explore Their Impact* by Neil F. Doherty and Heather Fulford, Loughborough University, UK, focuses on large Information Technology organisations in the UK. It presents a research framework designed to test whether and under what circumstances the adoption of an Internet Service Policy is likely to reduce the incidence of security breaches within large organisations.

Chapter XIII: *Metrics Based Security Assessment* by James E. Goldman and Vaughn R. Christie, Purdue University, USA, addresses the problem faced by organisations that, as a result of the September 2001 attacks on the USA, are investing more resources into security measures at the same time as they are enduring an increasing number and frequency of security breaches. The authors propose a method of measuring an organisation's information security.

Chapter XIV: *The Critical Role of Digital Rights Management Processes in the Context of the Digital Media Management Value Chain* by Margherita Pagani, Bocconi University, Milan, Italy, discusses the implementation of digital rights management by five media companies.

# References

Baird, R.M., Ramsower, R., & Rosenbaum, S.E. (Eds.). (2000).*Cyberethics.* Amherst, NY: Prometheus Books.

Day, K. (2003). *Inside the security mind: Making the tough decisions.* Upper Saddle River, NJ: Prentice Hall.

Edgar, S.L. (2003). *Morality and machines: Perspectives on computer ethics* (2nd ed.). Boston: Jones and Bartlett.

Gauntlett, A. (1999). *Net spies: Who's watching you on the Web?* Berkeley, CA: Frog Ltd.

Snow, C.P. (1964). *The two cultures and a second look.* London: Cambridge University Press.

Spinello, R. (2000). *Cyberethics: Morality and law in cyberspace.* Sudbury, MA: Jones and Bartlett.

Spinello, R.A., & Tavani, H.T. (Eds.). (2001). *Readings in CyberEthics.* Boston: Jones and Bartlett.

Williams, R. (1981). Communications technologies and social institutions. In R. Williams (Ed.), *Contact: Human communication and history.* London: Thames and Hudson.

# Acknowledgments

I would like to extend my thanks to Mehdi Khosrow-Pour for providing me with the opportunity to edit and contribute to this book and to the Idea Group staff for their guidance and assistance. My appreciation is also extended to the chapter reviewers whose efforts have contributed to the excellent standard of work in this collection. Most importantly, I wish to thank all of the contributors and reviewers, including Mr. George Kelley, and hope that you are pleased with the outcome.

Finally, I would like to express my gratitude to Associate Professor Kathy Blashki, Head of the School of Multimedia Systems, Monash University for her ongoing support of my research career.

*Dr. Marian Quigley*
*Berwick, Victoria, Australia*
*December 2003*

# Part I

Information
Ethics

## Chapter I

# MAMA on the Web:
## Ethical Considerations for Our Networked World

Barbara A. Schuldt
Southeastern Louisiana University, USA

## Abstract

*This chapter introduces ethical considerations that are especially relevant for the current networked world. It discusses the use of a mnemonic, MAMA — multicultural, adaptive, multifaceted, and archival — as a way to categorize ethical issues as we discover and discuss them today and in the future. By using these categories, the reader can evaluate how the Internet and, more specifically, the World Wide Web (Web) create new ethical concerns as information technology innovation and users drive new Web-based applications and discoveries. In addition, this chapter will pose key ethical questions that will help stimulate the reader to think about Web ethics. In thinking about these questions the reader will explore and hopefully discover his or her own past learned user behaviors and their potential for adverse ethical consequences to the individual and to society. It is through thinking and discussing the ethical consequences of Web-based applications that society will become aware of our own ethical norms and assess how we would respond before we electronically encounter ethical dilemmas.*

# Introduction

Information systems ethics, or the more popular term, *computer ethics*, have been studied for several decades (Bynum, 2000; Wiener, 1950). The study of information systems ethics has primarily focused on the impact information systems in general have had on society. This research has been very valuable and has helped information systems professionals and students assess various business scenarios using ethical constructs prior to these situations being encountered in the "real world". As a result of these studies we do have a strong sense of information systems ethics. However, the pervasive nature of the Web has changed who is involved in the use of information technology systems and the significance of the changes to the individual and society. To help us better understand the significance of the changes, a short history of the Internet to the current Web will follow.

# Background

The Internet was built in an open systems environment, where collaboration to advance the improvement was encouraged. But the Internet was a closed community of a few intellectuals, primarily in the United States of America. In this spirit many individuals spent countless hours expanding the capabilities of the Internet without worrying about ethical issues. As long as the Internet was used within this relatively closed community — government and academics — ethical problems could be discussed and resolved. Resolution was through the use of peer pressure and chastising the offending users. Even though Internet use has been extended to commercial venues, the software actually running the Web is non-commercial. The Internet commercial pressures exploded in 1991 with Tim Berners-Lee's application, the World Wide Web. With the Web, users now had virtually seamless interfaces through the use of browsers. Around 1995, the Web transferred from a government run network to a network run by commercial organizations. One of the first issues that had to be resolved concerned the limited assignment and control over domain names (DNS). After worldwide discussion the Internet Corporation for Assigned Names and Numbers (ICANN) was empowered to oversee the new DNS assignments. The first new registrars were announced in April 1999. Assignment of domain names was no longer a government sanctioned monopoly.

The Web community deals with other problems to ensure the continued safe use of the Web. Anonymity can be problematic, but has been resolved so anonymous emails of spammers can be identified and stopped. Anti-spam software can be used to minimize personal spam attacks to the individual. In the anonymous Web world, digital signatures have been a stumbling point to e-commerce. This has been resolved so that the digital signatures can be verified. Additionally, broadcasters use the Web to transmit video and audio along with static Web pages. Digital online music caused ethical as well as legal negotiations in recent years over the new format MP3. It should be noted that the distribution of print, audio, and video media through each new medium was problematic initially but then was resolved so that it can be done relatively seamlessly. For example, the introduction of the printing press created a need to resolve copyrights and a payment for the use and/or distribution of copyrighted material.

The Internet community is already planning for the next step—a larger Internet —with the adoption of Internet Protocol version 6 (IPv6), which increases the size and availability of IP addresses, and will allow more data and encryption in the packets transmitted on the Web. Advanced search engines are being developed that incorporate language translation and artificial intelligence to facilitate sophisticated Web searches. In addition, discussions are underway about expanding the global nature of the Internet and third world countries' access to this technology (Schenker, 2003). It is very difficult to keep up with all the new advances on the Web, which can happen very quickly. Therefore, there needs to be an identification of several basic categories to help us define and discuss ethical issues on the Web. Continued technological innovations will keep the Web a wonderful computer information network.

## Web Ethical Issues

Beyond the technical innovations are the creative individuals who see the Web as a global community with endless commercial and not-for-profit applications. This has led the Web to be the vehicle for the expansion of marketplaces and spaces beyond traditional restricted physical locations previously imposed on businesses. This marketing pressure, along with incessant new technology (computer hardware and software) introduction, has forced organizations to continually redefine their Web presence. During this redefinition, organizations should be taking the time to determine if their Web presence could create ethical tribulations for a diverse global community. However, this is usually only done after there are problems. And unfortunately, the problems are identified by dissatisfied and sometimes disgruntled users. So what should an organization's ethical considerations be when it is refining its Web presence? The subsequent sections will present these ethical considerations in a structured framework.

Organizations must take into consideration the ethical perceptions of many culturally diverse nationalities and languages when they begin or maintain their Web presence. The World Summit on the Information Society (WSIS) conference in December 2003 looked at the Internet policies worldwide. They noted a concern over rural areas where the information infrastructure is unaffordable or restricted by some national governments. This summit promotes a free exchange of ideas and knowledge that incorporates the ethical, legal, and sociocultural dimensions (WSIS, 2003). This is reflective of the Web moving towards a global electronic community with a single acceptable and hopefully ethical culture and language. It is unclear if this is being planned and discussed or if those who were unethical on the Web are being ostracized by the Web community, thereby minimizing their actions. According to Roger Darlington (2002), "increasingly the debate about the content of the Internet is not national but global, not by specialists but by the general populace. There is a real need for this debate to be stimulated and structured and for it to lead to 'solutions' which are focused, practical and urgent." The Web is no longer the domain of techno geeks or techies but is now used by all classes of individuals.

The tragic events on September 11, 2001 in Washington, D.C. and New York City demonstrated the pervasiveness of society's use of the Web to distribute and find out information worldwide (Pegoraro, 2001). News bureau Websites like CNN were jammed and access to this and similar sites was extremely slow. But for many, the Web was their only access to information about the 9-11 terrorist events, since many other news sources were disabled that day. This is an example of how the Web has been assimilated and integrated into our lives, since the information available on the Web was as close to "real time" as possible.

The number of individuals using the Web continues to grow exponentially. In January 2000 it was estimated that the worldwide Internet population was 242 million (CommerceNet, 2001). Commerce Net (2001) also forecasted that the Internet population would reach 765 million by the year-end of 2005. Much of this growth will be through the use of Web appliances or Internet-enabled appliances (Cerf, 2002). According to Vincent Cerf, (2002, p. 80) "we're already starting to see a great many automatic devices. … There is an Internet-enabled refrigerator made by Electrolux in Sweden. It has a liquid-crystal touch-sensitive display and is a nice addition to today's household communication, which is made up mostly of paper and magnets on the front of the refrigerator. You can use it to send e-mail and surf the Web." Sony Ericsson will, in the autumn of 2003, be bringing out a cell phone that includes the functions of a pocket PC, camera, and other features (Blecher, 2003). "Ericsson has estimated the number of mobile devices that will be Internet-enabled by 2006 to be 1.5 billion devices" (Cerf, 2002, p. 76). Along with the enhanced cell phones and Web appliances, research is being conducted on wearable computers. Fossil introduced a wristwatch PDA (Broersma, 2002). This adaptability of the Web has led to its assimilation into our everyday lives. The Internet has been adapting since its conception into the current Web we now use. Today's Web is a complex set of multiple networked computer systems that share and pass data, information and files around the world. The push to have higher speeds and expanding bandwidth pipes to increase the throughput of our data transmission throughout the Web has never been greater. We are demanding more from our networks and they are continually being adapted to accommodate our needs and desires.

In addition to being global and adaptable to everyday life, the Web must be multifaceted. The Web or, more specifically, the Internet, was used as a communication and file sharing system. This has evolved into the graphical interface Web of today that is used to communicate, to share data and information, and to assist in commercial ventures. Individuals and organizations are using the Web to promote and sell products and services. Today's Web is used to sell or promote individuals for political or other positions, as well as ideas, products, and services. In today's global business environment, it is the end result that is important. And the result is to promote the Web content worldwide. It is amazing the amount of material that is available on the Web. It is estimated that this material is doubling every 180 days. Yes, there is junk on the Web, but there is also quality. Vincent Cerf (2002, p. 78) observed that he continues "to be astonished not only at the amount of information that is on the Net but also at its quality". Later in the chapter, how we are using the Web will be demonstrated.

Since we have become dependent on the Web and, from all accounts, we will continue if not increase our dependence on the Web, it is important for the data and information to be available. The Web is a 24 * 7 * 365 ¼ technology. This means that any time and from anywhere a Website can be accessed. Websites must be timely and accessible, but for how long? The Web is still in its infancy and as such has not faced the necessity of defining archiving protocols. Older material is either removed from the Website by the creator or archived and accessible on the original or another Website, or the Website is no longer supported and you have a bad link. This has changed how we look at information systems and their long-term viability. From a societal perspective, we have an ethical issue concerning what, when, and how material should be archived.

This assimilation of the Web into the lives of billons of people has impacted and will continue to impact ethical decision-making. Connecting to the Web is not a one-way connection. If we are connected to the Web then the Web — the global online community — is connected to us. This was very clearly demonstrated in the recent spread of the MSBlast worm (Lemos, 2003). This two-way connection can not only increase the Web-based applications but also increase our need for a procedure to focus our thinking about the ethical scenarios. One procedure to help us solve ethical scenarios was published by Ernest Kallman and John Grillo in 1996. To determine if an ethical dilemma exists, their procedure has you apply several informal guidelines. These guidelines include:


- if you or others would prefer to keep the situation quiet,
- does it pass the Mom test (would you tell your mother and what would she do?),
- does it pass the TV test (would you tell a TV audience?),
- does it pass the Market test (can you advertise it?),
- does your instinct tell you something is wrong, and
- does it pass the smell test (Kallman & Grillo, 1996).


Now, in addition to these ethical tests, there should be the ethical "Web test". The "Web test" could have you evaluate whether you or your organization would put this on your Website and be able to state that the information is valid. The "Web test" must take into consideration the global nature of the Web and not be ethnocentric. This "Web test" will need to be very adaptive since the Web in some ways is like a living entity, growing, learning, and evolving. In addition, the "Web test" will need to allow for the multifaceted nature of the Web. It is many things to many people and the ethical concept should not preclude a Web application because it was not present during the original "Web test" definition stage. And the "Web test" must include an assessment of the long-term availability of the data or information, which is hard given that the Web is a complex set of networks without a central legislative or regulatory body. To help with our discussion about ethical decision-making, instead of an all-encompassing "Web test," four ethical constructs are presented. These four ethical constructs are: multicultural, adaptive, multifaceted, and archival.

In Richard Mason's article, "Four Ethical Issues of the Information Age" (1986, p. 486), he stated, "The question before us now is whether the kind of society being created is the one we want." He challenged the information systems professionals to develop ethical standards for the permeation of computers in our everyday life. He proposed the ethical issues of privacy, accuracy, property and accessibility (PAPA) as a way to focus the ethical discussion regarding information systems. While these four issues continue to be relevant today, the Web requires additional ethical considerations and discussions. No longer is it the information systems professionals who must develop ethical standards but all of us as users of the Web. To help the global electronic community focus the ethical discussions about the Web, this chapter proposes the following ethical categories — multicultural, adaptive, multifaceted, and archival. The acronym MAMA can be used to help us remember the Web's ethical issues relevant today. These four ethical issues are appropriate when we evaluate possible ethical scenarios that could occur from information systems that are networked via the Web. Identification of these concepts will hopefully lead us to answer Richard Mason's question (1986, p. 486) by being able to state that the society being created today is the one that we truly want. Our society today can be described by both offline and online aspects; however, the online aspects have infused our daily lives at an unprecedented rate.

In the next section, the four concepts will be defined and questions are suggested to help with the ethical considerations. Following the four definitions and questions the chapter will provide examples of ethical situations and how they would apply to the four concepts.

- *Multicultural* refers to the global nature and unifying movement towards the assimilation of many cultures into one Web culture. This Web culture includes values, norms, and problems arising from different languages. Questions to aid in ethical dialogues about actions on the Web impacting

the Web culture and more interestingly, our other non-electronic culture
are:

- What is a global culture?

- How is a global culture determined?

- Can we retain the ethnic diversity and uniqueness that contributes to
  the wonder of the world?

- Do country-specific values have a place in the global Web culture,
  such as free speech, equality and justice? Free speech, equality, and
  justice would be values that people of the United States of America
  would consider to be part of our culture.

- How are conflicting values handled?

- What about the digital divide (access and availability to information
  systems and technology) between the haves and have-nots?

- Is the Web helping bridge the digital divide or increasing it beyond
  repair?

- Should the Web's culture be planned for or allowed to evolve?

- *Adaptivity* is defined as the Web's ability to evolve as innovative new
  applications are developed and user needs change. The adaptation
  includes not only the computer hardware, software and networks but the
  individuals using the Web and the ability to transition to the new Web-
  based applications. Questions to help assess the consequences of these
  adaptations are:

  - Who should initiate new Web-based applications?

  - How can the Web community help individuals and organizations
    adapt to the changing Web environment?

  - What training or education must be provided for the adoption of the
    new applications?

  - Who provides the training or education? At what cost?

  - What are appropriate changes to the Web?

  - What changes are being implemented to prey on the unsuspecting
    user or users without training and education?

  - How does the adaptation affect those without access to the Web and
    those with older information systems and technology?

- *Multifaceted* references the broad use of the Web for many different applications. Many of these applications are conflicting and users are unaware of the consequences of their use of some applications. Questions to assist in evaluating the benefits and costs — both tangible and intangible — of a new application are:

  - What is the minimum that a Web presence should provide in terms of information, ownership, verifiable claims, or security?

  - How can users identify what to expect from the Web presence?

  - How can users evaluate claims made on the Web?

  - Are the individual's rights maintained?

  - How are legal problems from one nation resolved when the Web application although global is under the providence of another country?

- *Archival* refers to the need to have an established guide or process for archiving the data and information on the Web. Questions that facilitate a discussion on this concept are:

  - Should Web material be pulled off of the Web by an external entity — government, Internet service provider, and so forth?

  - What, when, how and who should archive the material on the Web?

  - Who should have access to the material in the archives and for how long?

  - How are broken links and Web pages that are no longer accessible treated by users?

The following section discusses possible ethical situations that demonstrate the four ethical concepts.

# Multicultural

The Web is a distributed network of networks, which does not have any one governing body for either regulatory control or legislative jurisdiction. From a historical perspective the roots of the Internet are with a few intellectuals primarily in the United States of America. Along with this came many of the values of the USA and western Europe. However, this is no longer the predominant population using the Web since only two-fifths of the users still reside in the United States. The Internet demand continues to grow at a factor of 1.4 to 2 times per year (Cerf, 2002). Much of this new demand is from Asian and African residents. CommerceNet (2001) forecasted that by the end of 2003, 60% of the world's online population would be outside the USA.

The ethical issue of multicultural looks at the cultural relativism of Websites. How does an organization ensure that its Website will be appropriate for the cultural values and ethics of the world? This is extremely difficult. What is appropriate in the USA may be offensive in Japan, in China, or in Egypt. How does the organization remain a respected global citizen if parts of the global electronic community question its Website's values or avoid its Website due to cultural conflicts? This is an issue that needs to be looked at by the global electronic community. The Web is the technology that has and will probably continue to facilitate the growth of the global electronic community. The question is, will it be many communities using the Web but not interconnecting due to value and cultural differences or will it be one global electronic community?

It appears that the Web by its global nature is establishing a Web culture that is a composite of the cultural norms of the participants. By default, are we defining this Web culture as acceptable since we have not discussed what should be an ethical Web culture? By allowing this to evolve and be controlled by those organizations that have the resources to establish and promote Websites, we may be missing or eliminating the very diversity of thought and contribution that built the Internet. Often we look to laws to follow regarding established ethical problems, but in the global community laws are nation-based and the Web has no predefined national boundaries.

Also, most USA citizens want this global community to be free of governmental influence, regulation, and control. Usually at the point the government becomes involved, it is to pass legislation that some would characterize as restrictive to the software application creativity that made the Web the persuasive influence in our lives today. A couple of recent examples of this type of user behavior that has ethical consequences are discussed. First, there is the Children's Internet Protection Act passed in 2000 by the USA government. The act was passed with good intentions but the legislative body failed to understand the nature of the technology. The filtering software that libraries were required to use would not only restrict pornographic sites but also sites on safe sex and health

(Nunberg, 2003). Another example of the USA-centric approach to the Internet comes from John Markoff and John Schwartz (2002), in their article for *The New York Times,* which critiques President Bush's administration's proposal for a system that would require Internet service providers to "help build a centralized system to enable broad monitoring of the Internet and, potentially, surveillance of its users". Historically, the USA was the predominate entity in the creation of the Internet, but that is no longer true.

Europe has been proposing similar action to monitor the Internet's traffic. This proposal would require that "ISPs and other network operators retain data on telecommunications usage, such as records of e-mail and Internet use, for seven years" (Meller, 2003). This is being criticized and challenged through the European legal system by various opponents. A more recent initiative came from the Council of Europe Convention on Cyber-Crime, which is seeking to stop "hate" speech on the Internet. Opponents of this initiative, including the American Civil Liberties Union and the U.S. Chamber of Commerce, fear that American individuals and businesses could be subjected to this proposal even if the United States of America does not agree or sign on with this initiative. "They warned that users could be coerced into adhering to other nations' hate-speech laws, while Internet providers might be forced to keep an eye on their customers for activities that violate such laws" (Reuters, 2002, February 6). In addition, Reuters (2002, December 10) reported that "Australia's highest court ruled Tuesday that a defamation case sparked by a story on a U.S. Website could be heard in Australia, opening a legal minefield for Web publishers over which libel laws they must follow. The landmark ruling that an article published by Dow Jones & Co was subject to Australian law — because it was downloaded in Australia — is being watched by media firms as it could set a precedent for other cases" (Reuters, 2002, December 10). It is feared that as a result of this ruling a backlog of lawsuits will emerge and subsequently entangle further technological development. The potential litigation process may preclude many from capitalizing on this ruling. Ideally, the international Web community should work toward a resolution and agreement about the issue of physical Web server location versus where material was downloaded. The international community did work together in the past to resolve the conflicts in domain name assignments.

It is predicted that China will become "the largest Internet user community in the coming decade, followed by India, which will become the second largest Internet user community toward the end of the decade" (Chon, 2001, p. 116). If China is the largest Internet user community, will that give them the clout to set the culture for the Web? Current Chinese rulers, like other Communist leaders, have retained their political power by controlling access to data and information (Vietnam's Cyberdissident, 2003). Recently, China allowed Internet cafes to open but under new rules that ban minors, require records be kept on customers and what they access and ban Websites deemed politically sensitive (Bodeen, 2002). Governmental control over media and access to information has been a common tactic in nations that fear losing control. The Web facilitates the transmission and availability of data and information. Moor (2001) claims that in the computer revolution, we have "recently entered the third and most important stage — the power stage — in which many of the most serious social, political, legal, and ethical questions involving information technology will present themselves on a large scale" (p. 89).

In addition to governmental control on the growth and accessibility of the Web, other emerging countries in Africa, the Middle East, and Central Asia may not benefit from the Web. These countries may in fact help broaden the digital gap between the haves and have-nots (Chon, 2001). According to Burn and Loch (2001), the major facilitator for globalization is information technology and underdeveloped regions will be able to have access to the worlds' resources and expertise, which should boost economic development. Opponents argue that only capital is mobile and those individuals will still be exploited by developed nations that are supported by the Internet. Burn and Loch (2001, p. 13) stated that, "the Internet, used creatively can serve to begin to reduce the growing and persistent gap between the haves and the have-nots".

In larger cities in the United States, ethnic targeted Websites have been successful. Examples of these Websites in Houston are *blackhouston.com, insideblackhouston.com,* and *thehoustoncalendar.com* (a Latino events calendar) (Guy, 2002). These Websites target various ethnic groups within a city and provide helpful information to these groups. In Cairo, a new Website, *Islam-Online.net,* was created to "present a positive view of the faith to non-Muslims, to strengthen unity in the Muslim world and to uphold principles of justice, freedom and human rights" (Wakin, 2002). These few Websites demonstrate the expression of diversity present on the Web. It will be important that, as we internationalize further, the ethnic differences and unique cultures are not lost.

This leads us to the need for the Web to adapt to changes from its very nature of promoting and encouraging innovative applications. This adaptation, however, has ethical issues.

# Adaptive

The Web has been and will continue to be very adaptive and responsive to new technologies and innovations in the field. However, this adaptive user behavior can have ethical concerns for many of its users. Such questions are:

- designing Websites for the lowest common capabilities of the target audience's computer hardware and software versus continually designing for the leading edge hardware and software;

- when new technologies should be adopted and incorporated into the organization's Web presence;

- who will assess the potential impact of these adaptations and evaluate their effectiveness; and

- how novice users will be made aware of their need to upgrade and improve their skills.

Or is this creating a society of haves and have-nots? Those that can access the Web any time and from anywhere have an advantage over the individual or group that has limited or controlled access to the information on the Web. The international court systems will in the future be dealing with the fairness issue with business transactions. Such questions are: when do transactions begin and when do they end, how do varying line speeds, noise, and interference modify or negate a transaction, and do limited or restricted access and the availability of information change how organizations do business on the Web? As noted in the multicultural section, the physical location of the Web server has been where countries have had their legal jurisdictions. The example from Australia, which allowed for legal action against a Website not found in Australia because the file was downloaded in Australia, could have major ramifications for the media and organizations with Websites. This will be an issue for the international legal system to officially resolve. For now, organizations must be cognizant of users' ability to access the Web and where potential users might reside.

It is very frustrating for users to work with a Website over time and then see a dynamic design shift. These shifts can take advantage of new computer hardware and software but can be detrimental to the user just trying to do his or her job. What appears to be lacking is consideration for the user; is there an assessment of the user's need for training and a determination of when training should take place? One of the principles of systems analysis is to design a system that facilitates the user doing his or her job. Is taking into consideration the potential user's level of expertise in using information technology a Website redesign firm's responsibility or is it the user's responsibility? Organizations that depend on their Web-sites being used by stakeholders in the organization know how important it is to have an easy to use Website for the broadest audience. In addition, assuming the organization does change the Web design, and the new design requires training, how does the user acquire this training? Is this an unfair (unjust) added burden (cost and time) to the user? This is an example of the struggle faced by individuals and organizations who want to use the state of the art technology but do not want to leave out any possible visitors, and who need to upgrade the Web presence because of the current dynamic business environment, where data and information do not remain static. One suggestion is to provide users several training options depending on their information technology and/or experience level, but this can be expensive to the Web development and support team.

Furthermore, how quickly can the global electronic community respond to new questionable activities, such as Gamespot's download manager hiding spyware and digital-rights-management (DRM) on your PC (Hachman, 2002)? This software is actually "stealware" that diverts commissions for online purchases that you make from other vendors back to the vendor that put the DRM on your computer, rather than to the vendor who referred the purchase. Many of these referral vendors are dependent on these commissions for their survival (Schwartz & Tedeschi, 2002). A discussion about this topic generated 31 pages of postings in one day on *Slashdot.org*. The discussions defined this as unethical user behavior, but basically felt that it was not unusual and the respondents felt powerless to stop this kind of behavior. How can the global electronic community keep the adaptive, responsive nature of the Web as a high priority, while discouraging or restricting negative activities such as what Gamespot recently did with the spyware and DRM?

Other examples of potential adaptive Web ethical concerns include:

- IP Telephony, which needs to address the issues of fair cost, quality of service, and accessibility (Garcia, 2001; Greenberg, 2003; Zachary, 2003);
- Web agents or spiders (Eichmann, 2002);

- monitoring and storing e-mail, instant messages and sites visited (Markoff, 2003); and

- customized Web ads and pricing differences based on customer profiles (Ives, 2003).

These diverse examples support the concept of a multifaceted Web environment.

The Web by its very design is trying to provide a venue for anyone to do many things by being networked to the world. Does it set up additional ethical concerns that one entity — the Web — is trying to be multifaceted?

# Multifaceted

There is no dispute that the Web has many aspects. As of February 2002 there were 544 million Internet users, whereas the telephone companies have about 1.3 billion telephone jacks in which to plug a telephone or other device that will use the telephone line (Cerf, 2002).   In the push for increased mobility the wireless telephone providers have been incorporating Web access and applications with the cell phone. It is estimated that by 2006 there will be 21.4 billion Internet devices along with other global system for mobile communication (GSM) and 3rd generation wireless communication devices (Cerf, 2002). It would be hard to believe that all these users would be accessing the Web for only one or two standard applications. Therefore the Web must continue to be multifaceted as it has been.

The purpose of a Website can be for education or informational sharing, or to promote an idea, product or service, or to actually facilitate exchanges between two or more parties. How can these Web-based information systems that have a global influence of unprecedented economic impact support the ethical standards that society expects?

One area is the future of business-to-consumer (B2C) Web-based applications. The average consumer is still uncertain about the security of these transactions, whereas the business-to- business (B2B) and business-to-government (B2G) applications have grown exponentially. The success of these applications has reduced costs and increased revenues amounting to millions of dollars. In 2006 the growth to 21.4 billion Internet devices will not be from new organizations but primarily from individual users (Cerf, 2002). The average consumer is beginning and will continue to demand additional applications that allow him or her the freedom to take full advantage of the Web while remaining in a secure environment. How do we define what is a secure environment? The court system uses what a reasonable person would expect in assessing whether a person's privacy has been violated; will that work in the Web environment?

We already have an Internet-enabled refrigerator made by Electrolux in Sweden (Cerf, 2002); can other Internet-enabled appliances, jewelry or clothing be far behind? How are these multifaceted applications going to change society? And will it be the changes that we as a society want?

On top of this are demands by some Web users for adult activities. How will we ethically resolve our predisposition (at least in the USA) to the notion of free speech and the right to pursue adult activities while protecting those not capable of making informed decisions (children or mentally challenged individuals)? Offline ethical standards do not easily translate into the online world. Adult content after 9 p.m. has no correlation on the Web since it is always after 9 p.m. somewhere in the world.

Currently, with the fast-paced growth of Web-based applications, organizations are faced with the problem of Internet techies who do not have the same concern for the non-technical user that the new Web marketers are trying to cultivate. The techie wants to develop the system for the sake of the technology and because he or she can. The Web marketer is creating a new market and building an Internet community for the product or service. This can lead to conflicts within the organization that are often ethically based and will need to be resolved.

Another example of the multifaceted nature of the Web comes from the People for Internet Responsibility (PFIR) in their on-going discussions of Internet idealism versus income. "Peter Neuman of SRI International and Lauren Weinstein of Vortex Technologies warn that what should have been a tool to benefit users everywhere is instead being turned into just another asset for self-interests" (Internet Battles, 2001).

When we look at the multifaceted nature of the Web it is evident that each of us has different reasons for using the Web and different expectations from our interactions on the Web. One of the most frustrating experiences is to not be able to access a Website where you previously found material. It is this archiving of Websites and material that can lead to ethical problems for many organizations.

# Archival

Archival questions need to be answered by organizations, such as what is archived; where is it archived (internal or external sites); what audiences have access rights to the archives; and how long must the organization's materials be archived? Why is this important to the user? According to Hauptman (1999) "there is, however, often no way to evaluate the validity, truth, or accuracy of much that passes for electronic information. We do not know if what we read was intended by the author, subsequently altered by the originator or a hacker, or sullied by inadvertent error" (p. 5). It has never been truer that when material is coming from a Website, garbage in leads to garbage out. An individual or organization must protect the integrity of its Website to build trust with the reader.

Organizations today are faced with new data storage rulings. Firms need to determine how to save information in a systematic manner. They need to determine whether to internally or externally store their e-mail and instant messages. "Everything from Martha Stewart to Enron has brought e-mail into focus. This is getting people's attention now and it's something you just have to do," says Jim Pirack, vice president of compliance at ShareBuilder Securities, a subsidiary of Netstock Corp. (Barrett, 2002). Granted, Mr. Pirack was not talking about Internet posted material specifically; however, corporate e-mail and instant messages and, in our current business climate, publicly posted material on the Web, will probably be viewed as fair game for future legal actions. The Securities & Exchange Commission (SEC) has regulations defining how organizations under SEC jurisdiction must archive their information.

The online magazine Salon removed an article that had charged Thomas E. White, secretary of the Army, as a contributor to the accounting practices that led to the downfall of Enron. The article was removed according to the editors because "a critical piece of evidence, an e-mail message attributed to Mr. White, could not be authenticated" (Carr, 2002). The article was online and publicly available; it has now been removed, but where is it archived and referenced (for example, this chapter)? This specific case will probably be resolved in the legal system. But what is the impact on the rest of the Web? Does archiving become an individual's or an organization's responsibility? Or, as a result of legal action, offensive Web material must be removed, which can take considerable time and effort (Kennedy, 2000).

History has the potential of being changed or modified. A recent article in *The New York Times* by D. Milbank (2003) noted that the "Bush administration has been using cyberspace to make some of its own cosmetic touch-ups to history". They have removed not only the original transcript of Andrew S. Natsio's statement "that the U.S. taxpayers would not have to pay more than $1.7 billion to reconstruct Iraq" but all links to it as well, and "…the liberal Center for American Progress discovered that this link [the 'What's New' page] had disappeared, too, as well as the Google 'cached' copies of the original page" (Milbank, 2003).

The other problem concerns not enough material being archived on the Web. An Internet Archive, funded by the Library of Congress, the Smithsonian Institution, the National Science Foundation, and Compaq Computers, will be updated every two months. The archive will use spiders or Web bots to find pages for the archive. One ethical issue is password protected or copyrighted material. Page owners can request that their pages not be in the archive, but they must be proactive in making this request (Internet Annals, 2001).

In the court case The New York Times Company v. Jonathan Tasini, the court ruled that rights to publish online are separate from other rights. The New York Times Company removed articles by freelancers from their Website rather than negotiate the rights to publish them online. This is viewed by the academic community with mixed emotions. For the academic faculty member using newspapers to do historical research, this means it will be necessary to travel to each city. However, for the lazy student, he or she will not be able to rely on online publications and will need to use hard copy to complete his/her research (Newspaper Databases, 2002).

After Littleton, Colorado, police agencies want to search for e-mail threats. Kim McCreery, a spokesperson for America Online, stated, "the company moves up to 56 million e-mail messages per day and, like all Internet service providers, it eventually erases those records. The time before the erasure varies, but 'she said' a day old record would still be available to investigators while a month-old message might be gone" (Dyer & Scott, 1999, p. B5).

Many individual Website owners address archiving on the Web; however, many of us use the Web as a research tool and having access to historical archives may be of importance in the future as the Web matures. Ethical considerations of who is responsible for archiving Web materials, when they are archived, how this can be documented and verified, and who has liability for inaccurately archived material will need to be discussed and a workable consensus achieved.

# Future Trends

This chapter posed key ethical questions to stimulate the reader to think about Web ethics. The purpose of these questions was to help the reader explore and discover his or her own past learned behaviors and their potential for adverse ethical consequences. It is through thinking and discussing the ethical consequences of Web-based applications that individuals and society become aware of our own ethical norms and assess how we would respond before we electronically encounter ethical dilemmas.

"The Internet is quantitatively and qualitatively different from all earlier human experience" (Langford, 1999, p. 65). Except for rural areas, citizens in underdeveloped countries, and individuals with some handicaps, the Web has been and will continue to be an important part of our lives. It has been through a number of adaptations already as it evolved from a network of a small, tight group of users that basically had very similar ethical standards to one that has been opened up to the entire world. New applications will continue to be developed. Some are for the good of the global Web community such as niche ethnic Websites and other valid informational sites, but some applications seem to have other agendas such as the distribution of the Code Red, slammer, and MSBlast viruses or the hacker that broke into the weather computer network and interrupted the weather forecast for the English Channel, which led to the loss of a ship at sea (Markoff, 1993). As many look to the next generation Internet or Internet 2 to increase speed and expand bandwidth, ethical concerns will still be an issue. How can past experiences and behaviors be transferred to the next killer app in an ethical manner? We need to have a method to think and discuss possible effects from the next killer app prior to its global installation. The MAMA construct will provide the focus for the Web ethics today and in the future.

# Conclusion

As the global electronic community identifies the Web's ethical issues and proceeds to define acceptable Web practices related to MAMA - multicultural, adaptive, multifaceted, and archival -the World Wide Web will be assimilated into our everyday lives beyond our current limited thinking, in an ethical, responsible manner.

# References

Barrett, L. (2002, September 16). On message: How companies store communications. *Baseline* [Online]. Available: http://www.baselinemag.com/

Blecher, J. (2003, March 4). *Sony Ericsson's new cell phones for 2003* [Online]. Available: http://reviews.cnet.com/

Bodeen, C. (2002, October 12). China's Internet cafes face stricter limits under new rules. *The* [Baton Rouge, LA] *Advocate,* 13A.

Broersma, M. (2002, April 20). *Fossil connects palm to wrist bone* [Online]. Available: http://news.com.com/2100-1040-895713.html

Burn, J., & Loch, K. (2001, October/December). The societal impact of the World Wide Web: Key challenges for the 21st Century. *Information Resources Management Journal, 14*(4), 4-14.

Bynum, T.W. (2000, June). The foundation of computer ethics. *Computers and Society, 30*(2), 6-13.

Carr, D. (2002, October 4). Web article is removed; flaws cited. *The New York Times* [Online]. Available: http://www.nytimes.com/

Cerf, V. (2002, September/October). Internet musings. *EDUCAUSE Review*, 75-84.

Chon, K. (2001, March). The future of the Internet digital divide. *Communications of the ACM, 44*(3), pp. 116-117.

CommerceNet. (2001). *Industry statistics: Worldwide Internet population* [Online]. Available: http://www.commercenet.com

Darlington, R. (2002). *Internet ethics: Oxymoron or orthodoxy?* [Online]. Available: http://www.rogerdarlington.co.uk/Internetethics.html

Dyer, S., & Scott, D. (1999, May 7). Police search for e-mail threats. *Duluth News-Tribune,* B5.

Eichmann, D. (2002). *Ethical Web agent.* Retrieved from the author at eichmann@rbse.jsc.nasa.gove on September 15, 2002.

Garcia, B.E. (2001, May 28). Internet pay phones appear in U.S. airports. *St. Paul Pioneer Press,* C2.

Greenberg, S. (2003, April 1). *Is it time for Net2phone?* [Online]. Available: http://cma.zdnet.com/

Guy, A., Jr. (2002, March 5). Ethnic diversity fuels success of niche Web sites. *Houston Chronicle,* D1.

Hachman, M. (2002). Gamespot's "Download Manager" hides spyware, DRM. *Extreme Tech* [Online]. Available: http://www.extremetech.com/

Hauptman, R. (1999). Ethics, information technology, and crisis. In L.J. Pourciau (Ed.), *Ethics and electronic information in the twenty-first century,* (pp. 1-8). West Lafayette, Indiana: Purdue University Press.

Internet annals erect billions of Web pages back in time. (2001, October 25). *Los Angeles Times* [Online]. Available: http://www.latimes.com/

Internet battle is idealism vs. income. (2001, April 19). *Los Angeles Times* [Online]. Available: http://www.latimes.com/

Ives, N. (2003, February 11). Marketers shift tactics on Web ads. *The New York Times* [Online]. Available: http://www.nytimes.com/

Kallman, E.A., & Grillo, J.P. (1996). Solving ethical dilemmas: A sample case exercise. *Ethical decision making and information technology* (pp. 33-56). New York: McGraw-Hill.

Kennedy, R. (2002, December). Difficulty in applying laws in cyberspace when comparing the physical location of the Web server and where the grounds for legal action may be justified. *Computers and Society,* 53-54.

Landford, D. (1999). Some ethical implications of today's Internet. In L.J. Pourciau (Ed.), *Ethics and electronic information in the twenty-first century* (pp. 65-75). West Lafayette, Indiana: Purdue University Press.

Lemos, R. (2003, August 13). *Slapdash monster roams the Net* [Online]. Available: http://news.com.com/

Markoff, J. (1993, October 24). Cited on 15. *The New York Times,* CXLIII, E7.

Markoff, J. (2003, July 7). A simpler, more personal key to protect online messages. *The New York Times* [Online]. Available: http://www.nytimes.com/

Markoff, J., & Schwartz, J. (2002, December 20). Bush administration to propose system for monitoring Internet. *The New York Times* [Online]. Available: http://www.nytimes.com/

Mason, R.O. (1986). Four ethical issues of the Information Age. *MIS Quarterly, 10*(1), 486-498.

Meller, P. (2003, March 5). Europe hacker laws could make protest a crime. *The New York Times* [Online]. Available: http://www.nytimes.com/

Milbank, D. (2003, December 18). White House Web scrubbing. *The New York Times* [Online]. Available: http://www.nytimes.com/

Moor, J.H. (2001). The future of computer ethics: You ain't seen nothin' yet! *Ethics and Information Technology, 3*(2), 89-91.

Newspaper databases have become unreliable and frustrating. (2002, January 25). *Chronicle of Higher Education* [Online]. Available: http://chronicle.com/infotech/

Nunberg, G. (2003, March 9). Computers in libraries make moral judgments, selectively. *The New York Times* [Online]. Available: http://www.nytimes.com/

Pegoraro, R. (2001, September 17). The Net proves comforting in a time of crisis. *St. Paul Pioneer Press,* D2.

Reuters. (2002, February 6). Groups oppose Europe limiting online hate speech. *The New York Times* [Online]. Available: http://www.nytimes.com/

Reuters. (2002, December 10). Ruling in Australia may have big impact on Web news sites. *The New York Times* [Online]. Available: http://www.nytimes.com/

Schenker, J.L. (2003, December 11). U.N. meeting debates software for poor nations. *The New York Times* [Online]. Available: http://www.nytimes.com/

Schwartz, J., & Tedeschi, B. (2002, September 27). New software quietly diverts sales commissions. *The New York Times* [Online]. http://www.nytimes.com/

Vietnam's cyberdissident. (2003, July 7). *The New York Times* [Online]. Available: http://www.nytimes.com/

Wakin, D.J. (2002, October 29). Online in Cairo, with news, views and 'Fatwa Corner'. *The New York Times* [Online]. Available: http://www.nytimes.com/

Wiener, N. (1950). *The human use of human beings: Cybernetics and society.* New York. Houghton Mifflin.

World Summit of the Information Systems. (2003). [Online]. Available: http://portal.unesco.org/

Zachary, G.P. (2003, July 7). Searching for a dial tone in Africa. *The New York Times* [Online]. Available: http://www.nytimes.com/

Chapter II

# Establishing the Human Dimension of the Digital Divide

Helen Partridge
Queensland University of Technology, Australia

## Abstract

*This chapter will explore the human dimension of the digital divide. It argues that existing digital divide research takes primarily a socio-economic perspective and that few studies have considered the social, psychological or cultural barriers that may contribute to digital inequality within community. This chapter will discuss an ongoing research project that explores the psychological factors that contribute to the digital divide. Using the Social Cognitive Theory, the research examines the Internet self-efficacy of Internet users and non-users in Brisbane, Australia and San Jose, California, USA. Developing a psychological perspective of the digital divide will expand current understanding of a phenomenon that has far reaching social and economic implications. It will allow a more precise understanding of what is and who represents the digital divide in community. Organisations who are involved in bridging the digital divide will be better placed to develop strategies and programs that can more effectively narrow the gap between ICT "haves" and "have-nots".*

# Introduction

The digital divide between Information and Communication Technology (ICT) "haves" and "have-nots" has been a topic of considerable discussion since the U.S. federal government released its 1995 report on household access to technologies such as the telephone, computers and the Internet (NTIA, 1995). Since this time many organizations have endeavoured to bridge the digital divide through a diverse range of initiatives and projects. These initiatives and projects have been developed based on the current understanding of the digital divide. This understanding has been developed primarily from a socio-economic perspective. According to current studies (Lenhart, Horrigan, Ranie, Allen, Boyce, Madden & O'Grady, 2003; NOIE, 2002; NTIA, 2002), the primary factors contributing to the digital divide are income, employment and education. As personal computer prices have fallen and Internet services to the household are becoming increasingly less expensive, the socio-economic perspective of the digital divide becomes less convincing to explain all reasons for ICT non-use. The 1999 study by the National Telecommunications and Information Administration (NTIA) into the digital divide in the Unites States suggested that the "don't want it" attitude is fast rivaling cost as a factor explaining non-use of the Internet. Further support for this suggestion was more recently given by a Pew Internet and American Life Project (Lenhart et al., 2003) study, which stated that nearly one-quarter of Americans are "truly disconnected," having no direct or indirect experience with the Internet, whilst another 20% of Americans were "Net evaders," that is, people who live with someone who uses the Internet from home. Net evaders might "use" the Internet by having others send and receive e-mail or do online searchers for information for them. Recent criticism of the current digital divide studies (Jung, Qiu & Kim, 2001) has suggested that the studies fail to consider the psychological, social and cultural barriers to the digital divide. If all members of community are to be allowed to become active citizens and if community organisations are to develop services and resources that will contribute to bridging the digital divide, efforts must be made to more clearly understand the social, psychological and cultural differences that contribute to its development.

This chapter discusses a current research project into the psychological barriers of the digital divide. The chapter is divided into three parts. Part one considers what the digital divide is. A brief picture of the digital inequality in Australia and the United States is outlined. The limitations of current digital divide studies are discussed. So too is the relationship between information ethics and digital inequality in the information age. Part two outlines the current research project. The research approach, the underlying theoretical framework and the expected outcomes are discussed. Part three will discuss the future and emerging trends of digital divide research, suggesting further opportunities for study and exploration.

# The Digital Divide:
## A Review of the Literature

This section will provide a brief overview of the current digital divide literature. In particular, the section will discuss current understanding of the phenomenon based upon existing research aimed at quantifying and defining the divide, and considers the question of whether there are two digital divides. The section will examine current literature examining the psychology of the divide and will explore the relationship between the digital divide, information ethics and the information society. The section will finish with a brief description of the current digital divide challenge that the current research project is seeking to meet.

## Defining and Quantifying the Divide

The phrase *digital divide* has become the accepted manner for referring to "the social implication of unequal access of some sectors of community to Information and Communication Technology [ICT] and the acquisition of necessary skills" (Foster, 2000, p. 445). The term has been derived from the commonly held belief that access to Information and Communication Technology (ICT) such as the Internet, and the ability to use this technology is necessary for members of community if they are to fully participate in economic, political and social life.

Studies examining the digital divide abound. Three recent studies have been conducted in the United States (Lenhart et al., 2003; NTIA, 2002) and Australia (NOIE, 2002). Each study sought to establish a statistical snapshot of the current state of their nation's involvement with technology such as the Internet and computers. In the second of the studies in the U.S., the NTIA acknowledged that the Digital Divide "is now one of America's leading economic and civil rights issues" (NTIA, 1999, p. xiii). This statement is no less true for Australia. The findings from both the U.S. and the Australian studies highlight several interlocking factors that heighten the digital divide: race and ethnicity, geography, income, education level, employment status and physical disability. Individuals who can be identified through these factors are more likely to represent the "have-nots" in the digital divide.

## Two Digital Divides?

Several commentaries have emerged in recent years discussing the current studies measuring and quantifying the Digital Divide. In 2001 Jung, Qui and Kim considered the question, "What is the Digital Divide? Does it mean mere ownership of Internet connections…or does the digital divide describe more fundamental inequalities in people's connection to communication technologies?" (2001, p. 3). In considering this question the authors suggested that the current studies exploring the digital divide were limited by their focus on three primary measuring techniques. These techniques include: a *dichotomous comparison,* which focuses on the issue of simple access or ownership (i.e., computer owner vs. non owner); a *time based measure,* where more time spent online is equated to "regular use;" and a *measure of activities conducted online,* where frequency of engaging in activities such as online banking and online shopping are measured. Jung, Qiu and Kim suggest that these measures fail to consider the social context in which people incorporate technology. The authors suggest that the personal and social effects of the Internet must be considered in comprehending the more subtle aspects of the digital divide. Jung, Qiu and Kim suggest that once people have access to the Internet, the questions to be addressed are how can and do they construct meaning from their being connected. They conclude, "existing inequalities even after gaining access to the Internet can directly affect the capacity and the desire of people to utilise their connections for purposes of social mobility" (Jung, Qiu & Kim, 2001, p. 8).

Vernon Harper (n.d.) in a recent discussion paper suggests the existence of two digital divides: Access Digital Divide (ADD) and Social Digital Divide (SDD). The Access Digital Divide is based upon cost factors and is frequently discussed in terms of the presences of computers or Internet access in the household. The Social Digital Divide is "a product of differences that are based on perception, culture and interpersonal relationships that contribute to the gap in computer and Internet penetration" (Harper, n.d, p. 4). Harper recommends that the scholarly community build research that explores the social, psychological and cultural differences that contribute to the Social Digital Divide. Harper concludes by stating, "the issues surrounding the digital divide must be redefined away from the hardware and towards humanity" (n.d., p. 5). In agreement is Soraj Hongladarom (n.d.), who stated, "one should more accurately talk about the digital divide*s*, as there are many different kinds of the divide" (p. 3). In stating this, Hongladarom points to the work by Hargittair (2002), who argues the existence of a second-level digital divide that involves the gap between the skills people have when they are online. He contrasts this to the usual interpretation of the digital divide as little more than the gap between those who possess or do not possess the technology. The need to focus on an individual's ability to use technology instead of just accessing it is further explored in the recent work by Kvasny. In her 2002 doctoral dissertation, Kvasny undertakes a study of the cultural dimensions that contribute to digital divide in the United States. Kvasny suggests that her study "goes beyond describing the digital divide to analyzing *digital inequality*" (2002, p. 16). Kvasny uses the concept of *digital inequality* "to signify a shift and distinction in focus from *access* to *use* of information and technology" (2002, p. 16).

## The Social Digital Divide: Establishing a Psychological Perspective

Very little research to-date has attempted to explore the psychological factors that contribute to digital inequality. One of the first studies examining the psychology of the digital divide was undertaken at Michigan State University. Conducted by Eastin and LaRose (2000), the study examines the digital divide from the perspective of Bandura's Social Cognitive Theory (1986). This theory postulates that a person will act according to their perceived capabilities and the anticipated consequences of their actions. Self-efficacy is the primary component of the theory. It is the belief that a person has that they can perform a particular behaviour or task. Eastin and LaRose developed and validated an Internet Self-Efficacy Scale for the purposes of their study. Using university students, the study findings indicate that self-efficacy is a significant predictor of Internet use. According to the current socio-economic perspective of the digital divide, an individual with higher levels of education (i.e., university) is less likely to represent the digital divide than those with lower levels of education. Given that the study's participants were college students and no demographic data are provided, any conclusions drawn about the Internet self-efficacy and its role in the digital divide can be suggestive at best.

In the same year a study exploring the computer self-efficacy of African American high school students was undertaken by Foster (2001). Socio-economic studies of the digital divide suggest that African Americans are more likely to represent the digital divide. The study's findings suggest that African American high school students have a lower computer self-efficacy than non-African American students. Social and cultural factors unique to the study's participants suggest that the findings are not easily generalisable to the wider population.

Both studies expand current understanding of the psychological factors that impact upon a person's willingness to engage with information and communication technology. However, because of the participants used (i.e., college students and African American high school students), the studies can shed only limited light onto the impact Internet self-efficacy has on the digital divide within community.

## Information Society, Information Ethics and the Digital Divide

We live in a society where information is fundamental to the workings of everyday life. Everyone needs and uses information, whether buying a new car, opening a bank account, or writing a business report. We also live in a digital age where more and more information and communication is taking place in the digital or electronic environment. The Internet is the most public face of the emerging digital age. It is rapidly becoming the primary vehicle for information exchange and communications and is establishing itself as a vital and dynamic part of modern society. Governments at all levels use the Internet to disseminate information. Business and retail industries are flourishing on the Internet. The Internet provides a new and more flexible means of shopping for consumers. Learning and teaching is enhanced by the many resources and services offered via the Internet. The Internet also has a great impact on scholarly research and development. Sports fans can access up to the minute game results and view matches as they take place live in any part of the world. Keeping in touch with distant family and friends has become even easier with chat rooms and e-mail. As the Internet increasingly becomes a primary distribution vehicle for information exchange and communication, access to the Internet is no longer a luxury but a necessity for living in the current information age.

In 1998, Stichler and Hauptman asserted that the "information age has been widely acclaimed as a great benefit for humanity, but the massive global change it is producing brings new ethical dilemmas" (p. 1). In agreement is Luciano Floridi, who in a 2001 paper based on an invited address to the UNESCO World Commission on the Ethics of Scientific Knowledge and Technology (COMEST), stated that "the information society...poses fundamental ethical problems whose complexity and global dimensions are rapidly evolving" (2001a, p. 1). Floridi argues that "how information and communication technologies can contribute to the sustainable development of an equitable society is one of the most crucial global issues of our time" (Floridi, 2001b, p. 2). Floridi points to the digital divide in particular as the source of many of the ethical problems emerging from the evolution of the information society. The digital divide "disempowers, discriminates and generates dependency. It can engender new forms of colonialism and apartheid that must be prevented, opposed and ultimately eradicated" (Floridi, 2001a, p. 3). Floridi concedes that on a global scale the issues of health, education and the acceptance of elementary human rights should be among "humanity's foremost priorities" (2001a, p. 2); however, Floridi argues that "underestimating the importance of the [digital divide], and hence letting it widen, means exacerbating these problems as well" (2001a, p. 2). Floridi concludes by announcing that "our challenge is to build an information society for all," and this is a "historical opportunity we cannot afford to miss" (Floridi, 2001a, p. 4).

## Defining the Digital Divide Challenge

The community is rapidly being divided into those who are information rich — the "haves" and those who are information poor — the "have-nots". Steps must be taken to ensure that all members of community have access to and the ability to effectively utilize information and communication technology such as the Internet. By taking steps we will assist in preventing the creation of a digital divide, and ensure that all members of society have an equal chance of establishing and maintaining productive personal and professional lives. The digital divide is a complex phenomenon. Many studies to-date have taken the socio-economic perspective of the digital divide, in which income, employment and education are the primary factors influencing the development and growth of the digital divide. Whilst these studies provide a valid and important understanding of the phenomenon, the studies represent only a single layer of understanding to digital inequality (see Figure 1).

*Figure 1. The single layered perspective to digital inequality*

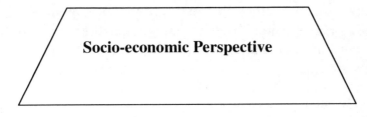

The psychological, social and cultural factors that may contribute to the digital divide are only now just beginning to be explored. Further studies exploring the psychological, social and cultural factors that prevent individuals from embracing technology in their lives would help in providing a more detailed understanding of the digital divide in society. Efforts that are made to more clearly understand the socio-psychological and cultural differences that contribute to the digital divide will ultimately ensure that all individuals have the opportunity to become active community citizens and allow organizations involved in establishing programs and initiatives to do so with greater efficiency and effectiveness. If adequate steps are to be taken to bridge the growing information and technology gap, a thorough understanding of the digital divide is vital. New layers of understanding need to be added to the current single layered perspective of digital inequality. The research project outlined in this chapter will contribute to this challenge by shedding light on the psychological factors contributing to digital inequality within communities. Figure 2 represents the multi-layered understanding of digital inequality that the current research (and others like it) will develop.

*Figure 2. The multi-layered approach to digital inequality*

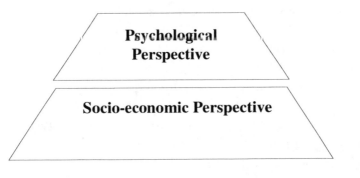

# The Research Project

## The Research Aim

The research project explores the human dimension of digital inequality by examining the psychological factors that contribute to digital divide. The research is focused by the basic question: Are there internal forces causing members of community to choose not to integrate information and communication technology, such as the Internet, into their lives? The main aim of the research is to explore the notion of the Social Digital Divide proposed by Harper (n.d.) by examining the Internet self-efficacy of Internet users and non-users within community. This will be achieved by:

1.    Measuring the Internet self-efficacy of Internet users and non-users.
2.    Determining if there is a difference in Internet self-efficacy between:
    •    Internet non-users users who represent the "Socio-economic Digital Divide" or as proposed by Harper (n.d.), the Access Digital Divide;
    •    Internet non-users who do not represent the Socio-economic Digital Divide but who may represent the Social Digital Divide as proposed by Harper (n.d.);
    •    Internet users who are not considered to be part of the digital divide.

## Theoretical Framework

The research design and data gathering techniques used in the current study developed as a direct result of the theory being explored. As such, developing an understanding of the theory being examined — the *what* — will help in understanding the research design and data gathering techniques — the *how* — being used within the current study. A brief outline of the theory used to form the framework for the current research will be provided.

## Social Cognitive Theory

This research will examine the internal or psychological forces that motivate an individual to refrain from integrating technology, such as the Internet, into his or

*Figure 3. The triadic relationship*

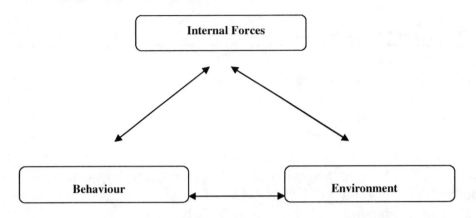

her life. To achieve this end the research will use the Social Cognitive Theory (SCT) developed by Albert Bandura (1986). This theory asserts that behaviour is best understood in terms of a *triadic reciprocality* (Bandura, 1986), where behaviour, cognition and the environment exist in a reciprocal relationship and are thereby influenced or are determined by each other. According to Bandura, individuals are actively involved in shaping their environments and not merely passive reactors to them (Bandura, 1986). This relationship is shown in Figure 3.

Social Cognitive Theory makes the following assumptions:

- People have powerful *symbolising* capabilities. Through the formation of symbols, such as images or words, people are able to give meaning, form and continuity to their experiences. In addition, through the creation of symbols people can store information in their memories that can be used to guide future behaviour (Bandura, 1986).

- People can *learn vicariously* by observing other people's behaviour and its consequences. This allows people to avoid trial-and-error learning and allows for the rapid development of complex skills (Bandura, 1986).

- People are *self-reflective* and capable of analysing and evaluating their own thoughts and experiences. Such capabilities allow for self-control of thought and behaviour (Bandura, 1986).

- People are capable of *self-regulation* by having control over their own thoughts, feelings, motivation and actions. Self-regulated behaviour is initiated, monitored and evaluated by the individual to accomplish his or her own goals (Bandura, 1986).

- People's behaviour is directed toward particular *goals* or purpose and is guided by *forethought,* where forethought is people's capability to motivate themselves and guide their own actions (Bandura, 1986).

Whilst the Social Cognitive Theory consists of both the outcome expectancy construct and the self-efficacy construct, it is the latter which has been repeatedly shown in studies to be the central factor in determining an individual's choice to undertake a behaviour or task. This research will focus on self-efficacy as a predictor of an individual's choice to access and use technology.

## Self-Efficacy

Self-efficacy is a major component of the Social Cognitive Theory. Bandura (1986) describes self-efficacy as "people's judgments of their capabilities to organise and execute courses of action required to attain designated types of performances" (p. 391). Or more simply stated, self-efficacy is the belief a person has about his or her capabilities to successfully perform a particular behaviour or task (Cassidy & Eachus, 2000, p. 1).

Self-efficacy has three dimensions: magnitude, strength and generality (Bandura, 1986). *Self-efficacy magnitude* refers to the level of difficulty a person believes he or she is capable of performing (Maddux, 2000). *Self-efficacy strength* refers to the level of conviction a person has that he or she can perform a task or behaviour. *Self-efficacy generality* refers to the extent to which a person's success or failure in a task or behaviour will influence the person's self-efficacy in other tasks or behaviours.

According to Bandura (1986), individuals acquire information about their personal self-efficacy from six primary sources: (a) actual experiences, (b) vicarious experiences, (c) verbal persuasion, (d) physiological states, (e) imaginal experiences and (f) distal and proximal sources (Bandura, 1986). An individual's own performances, especially past successes and failures, offer the most reliable source for assessing self-efficacy (Bandura, 1986).

# Research Context

The research project has an international context. The two cities of Brisbane, Australia and San Jose, United States of America provide the communities through which study participants will be obtained. A brief profile of each city will follow:

- *Brisbane* is the capital city of the state of Queensland. It is the third largest city in Australia, covering an area approximately 1350km² and supporting a total population of 1,601,417. In the recently released planning document "Living in Brisbane 2010," the Brisbane City Council articulate its vision for Brisbane as a "smart city [that] actively embraces new technologies...Brisbane should seek to be a more open society where technology makes it easier for people to have their say, gain access to services and to stay in touch with what is happening around them, simply and cheaply. All residents will have access to the Internet, and the ability to use it." (BCC, 2001)

- *San Jose* is a city in the state of California. It is the third largest city in California, covering an area of approximately 176.6 square miles and supporting a total population of 918,800. It is the 11th largest city in the United States of America. San Jose residents speak more than 46 different languages. Nearly one-third of the city's population has a Hispanic background (30%) and just over one quarter (27%) is from an Asian background. The average household income is $46,000 and the population is mostly aged between 18 and 64 years (66%). The San Jose City Council has adopted a "smart growth" strategy to the future development of the city, where "smart growth" refers to a city that "feels safe...where new housing is affordable... where there are jobs for all its residents" (City of San Jose, 2001, p. 6).

## Research Approach

Self-administered surveys will be used in data gathering. Participants are Internet users and non-users from Brisbane, Australia and San Jose, United States. The survey instrument will consist of three sections: The first section seeks information on *demographic* details such as gender, age, employment status, income level and education level. The second section gathers data on the participants' *Internet use*. Data gathered included where they obtain access to the Internet, length of involvement with the Internet, self-perception of Internet skill and frequency of Internet use. The third section will gather data on the participants' level of *Internet self-efficacy*. Data collection will take place in 2004, with the results and key findings being available in 2005.

## Data Gathering Context

Public libraries in both the U.S. and Australia have invested large amounts of time, money and energy into establishing programmes and activities that will assist in bridging the digital divide within communities. There are over 1500 public library facilities in Australia located in rural, inner city, suburban and remote areas (ABS, 2001). In Australia, the public library is the most visited cultural venue, with 99.4 million visits during 1999-2000 (ABS, 2001). Almost 50% of the Australian population are members of the public library network (ABS, 2001). While statistics on library membership and use for the U.S. situation was not available, anecdotal evidence suggests a similar level of community support.

The commitment demonstrated by the public library network in both the U.S. and Australia in providing support to all members of community to access and use the information technology, such as the Internet, and the obvious support and use by members of the Australian community suggests that a public library is a logical starting point to access study participants. A brief profile of the two library systems used in the current research will follow:

- The *Brisbane City Council (BCC) Library Service* consists of 32 static branch libraries. It has a membership base of approximately 362,000 people, and serves a total population of 865,000, spanning approximately 1350 km². It circulates over 9 million items each year, and has over 4 million visitors during this time. The library service is a vital and active part of the community, offering a wide range of services and resources, including Internet access and training. Two branches of the library service will be used in this study - Inala and Indooroopilly. The branches were selected based upon their potential to provide access to participants matching the desired profile. Both library branches hold regular "Introduction to the Internet" training sessions for members of the public. The BCC Library Service was chosen for two reasons. Firstly, as the largest public library service in Australia, its large client base will help in obtaining a suitable sample for the study. And secondly, the Brisbane City Council's and the library's support for the philosophy of equitable access to technology and communication for all members of the Brisbane community will allow for smooth administration of the study. The support of the philosophy is clearly expressed in the Council's 2010 planning document (BCC, 2001).

- The *San Jose Public (SJP) Library Service* consists of 18 static branch libraries. It has a membership base of approximately 650,000 people, and serves a total population of 918,800, spanning approximately 177 square miles. It circulates over 9 million items each year. The library service is a vital and active part of the community, offering a wide range of services and resources, including Internet access and training. Four branches of the library service will be used in this study — Biblioteca, Amalden, Hillview and Dr. Martin Luther King Jr. The branches were selected based upon their potential to provide access to participants matching the desired profile. The SJP Library Service was chosen for two reasons. Firstly, serving a community of similar profile to that serviced by the BCC Library Service, the SJP Library Service will provide an interesting and applicable international comparison. And secondly, the San Jose Public Library's support for the philosophy of equitable access to technology and communication for all members of the San Jose community will allow for smooth administration of the study. This can be seen by the willingness and enthusiasm to support an Australian based study.

## Research Participants

Participants in the study will be Internet users and non-users from two communities — Brisbane, Australia and San Jose, United States of America.

Studies measuring and quantifying the digital divide from the socio-economic perspective have identified several interlocking factors that heighten the digital divide: race and ethnicity, geography, income, education level, employment status and physical disability. The studies argue that anyone who can be identified through these factors is more likely to represent the "have-nots" in the digital divide. However, these studies fail to consider the psychological, social or cultural barriers that may be impacting upon an individual's choice to access and use technology. The current profile of who represents the digital divide is not complete. According to Harper (n.d.), the socio-economic studies of the digital divide have merely developed a profile of the Access Digital Divide.

*Figure 4. Two digital divides: Access digital divide & social digital divide*

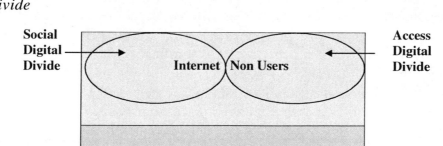

This research will help to expand the current profile of the digital divide by exploring the Social Digital Divide using the self-efficacy construct. To achieve this it is important that the current study uses participants who are (i) Internet non-users who represent the socio-economic view of the digital divide, or as proposed by Harper (n.d.), the Access Digital Divide; and (ii) Internet non-users who do not represent the socio-economic view of the digital divide, or as proposed by Harper (n.d.), may represent the Social Digital Divide. By ensuring both types of participants are present, the current study will be able to more fully develop a profile of the gap between information "haves" and "have-nots". More specifically, the study will be able to determine (i) if there are individuals who represent the Access Digital Divide but do not represent the Social Digital Divide; (ii) if there are individuals who represent both the Access Digital Divide and the Social Digital Divide; and (iii) if there are individuals who represent the Social Digital Divide but do not represent the Access Digital Divide. To provide comparative data the Internet self-efficacy of Internet users will also be examined. Figure 4 provides a graphical representation of the desired sample profile.

## Internet Self-Efficacy: A Definition

According to Bandura, there is "no all-purpose measure of perceived self efficacy" (Bandura, 2001, p. 1). Self efficacy scales "must be tailored to the particular domains of functioning that are the object of interest" (Bandura, 2001, p. 1). For the current research therefore to proceed, a clear understanding of "Internet self-efficacy" must be established. In 2000, Eastin and LaRose defined Internet self efficacy as the "belief in one's capabilities to organize and execute course of Internet actions required given attainments" (p. 1). Torkzadeh and Van Dyke (2001) suggest that Internet self-efficacy is the "self-perception held by individuals of their ability to interact with the Internet" (2001, p. 275). But what are "Internet actions"? What does it mean to "interact with the Internet"? Earlier this year the Pew Internet and American Life Project released findings from their most recent study of the use of the Internet by American citizens (Lenhart et al., 2003). Using 3,553 Americans, the study concluded that about 72 million American adults go online per day. According to the study the five most common daily activities performed by these Americans on the Internet include: sending e-mail, getting news, surfing the Web for fun, looking for information on a hobby and sending an instant message (Lenhart et al., 2003).

## Internet Self-Efficacy Scales

Four measurements of Internet self-efficacy existed at the commencement of the current research project. Details of each are provided below:

1.  *Internet Self-Efficacy Scale* developed by Eastin and LaRose (2000) is an eight-item self-report measure developed to assess an individual's confidence that he or she could use the Internet. In the scale, 7 corresponds to "strongly agree" and 1 corresponds to "strongly disagree". The higher the score obtained by an individual, the higher his or her level of Internet self-efficacy. Or more simply stated, individuals with higher scores possess a higher level of confidence in their ability to use the Internet. The scale has high internal consistency, with a Cronbach alpha of .93.

2.  *The Internet Self-Efficacy Scale* created by Torkzadeh and Van Dyke (2001) is a 17 item scale designed to "measure an individual's perception and self-competency with the Internet" (p. 1). The instrument is a five-point likert scale, where 1 corresponds to "strongly disagree" and 5 corresponds to "strongly disagree". The scale was developed using 277 undergraduates from a university in the southwest of the United States. The mean age of the student participants was 24.88. The scale has high internal consistency, with a Cronbach alpha of 0.96.

3. *The Computer/Self-Efficacy Scale* created by Maitland (1996) was designed to "take the computer self-efficacy scale closer to the 21st century…by including items measuring Internet self-efficacy" (1996, p. 1). More specifically, the scale is a measure of World Wide Web self-efficacy, as it focuses only on this aspect of the Internet. The scale was developed using 36 undergraduate students from a North American university. No data regarding reliability or validity are available. Because this scale incorporates both computer and Web self-efficacy, only those questions relating to the Web were included in the current survey instrument.

4. The Internet self-efficacy scale created by Hnilo (1997) was also located but a full copy of the instrument could not be obtained. As such the scale was not tested in the current study. The scale was developed using 37 students from a large midwestern university. The instrument was developed with a five-point Likert scale. The scale has good internal consistency, with a Cronbach alpha of 0.63.

Each of these scales was pre-tested on the target population (n=19). All the scales tested were developed using American college students. This research is the first time that the scales are being used with members of the community. Feedback from the research participants suggests that the scales were not appropriate for use with the current population. Many of the participants stated that they were unsure of some of the words used, including: "Internet hardware," "Internet software," "Internet program," "decrypting," "encrypting," "downloading," "URL" and "hypertext". Three of the participants did not fully complete the research instrument because they felt that the survey was "irrelevant to them" and "beyond their understanding". The three participants were all Internet non-users and therefore of direct relevance to the research. The preliminary results suggest that none of the existing Internet self-efficacy scales are applicable for use with community members. College students, even those who perceive themselves as Internet non-users or novices, may have a more highly developed knowledge and understanding of the Internet than members of community. Bandura recommends that measures of self-efficacy must be tailored to meet the specific "reading level" (2001, p. 4) of the population being examined. Consequently, it may be suggested that the "reading level" of participants within the current study (i.e., members of the general public) is significantly different from the "reading level" of the participants used in developing existing Internet self-efficacy scales (i.e., American college students).  The key finding was that an Internet self-efficacy scale for use within a community would need to be constructed. A brief description of this process will follow.

## Constructing an Internet Self-Efficacy Scale for Community

Researchers should follow a prescribed protocol in the development and testing of survey instruments in order to ensure the results are (i) reliable and valid in representing the domain of interest, and (ii) are not artefacts of the measurement instrument (Sethi & King, 1991). A protocol for developing self-efficacy scales has been well established by Albert Bandura (1986; 2001). This protocol is outlined in his seminal publication *Self-efficacy: The Exercise of Control* (1986) and more recent publication *Guide for Constructing Self Efficacy Scales* (2001). The current study will use this protocol in guiding the development of the self-efficacy measure.  Some of the key recommendations outlined by Bandura in constructing a measure of self-efficacy include:

- Self-efficacy is not a measure of personal skill. It is a measure of the belief an individual has regarding the use of personal skill under different sets of conditions (Bandura, 1986, p. 37).
- Self-efficacy should be measured at different levels of performance. "Sensitive measures of efficacy beliefs link operative capabilities to levels of challenge in particular domains of functioning," (Bandura, 1986, p. 38).
- Self-efficacy measures must be tailored to the particular domain of functioning that is being explored. "There is no all-purpose measure of perceived self-efficacy. The 'one-measure-fits-all' approach usually has limited explanatory and predictive value because most of the items in an all-purpose measure may have little or no relevance to the selected domain of functioning." (Bandura, 2001, p. 1).
- To achieve explanatory and predictive power, measures of personal efficacy must be tailored to domains of functioning and must represent graduations of task demands within those domains. This requires clear definition of the activity domain of interest and a good conceptual analysis of its different facets, the types of capabilities it calls upon, the range of situations in which these capabilities might be applied (Bandura, 1986, p. 42).

- The three dimensions of self-efficacy should all be considered when designing self-efficacy measures:
  - The *level of self-efficacy* belief should be assessed by the inclusion of items that represent varying degrees of challenge to successful performance within the domain of functioning. "Sufficient impediments and challenges should be built into efficacy items to avoid ceiling effect." (Bandura, 1986, p. 43)
  - The *generality of self-efficacy* should be assessed by the inclusion of items that measure self-efficacy in a variety of situations and contexts related to the domain of functioning (Bandura, 1986, p. 43).
  - The *strength of self-efficacy* should be assessed by asking participants to rate the strength of their belief in their ability to execute the requisite activity on a scale. Bandura recommends a 100-point rating scale ranging in 10 unit intervals from 0 ("Cannot do") through escalating degrees of assurance; 50 ("Moderately can do"); to complete assurance, 100 ("Certain can do"). Efficacy scales are unipolar, ranging from 0 to a maximum strength. They do not include negative numbers because a judgment of complete incapability (0) has no lower gradations. Scales that only use a few steps should be avoided because they are less sensitive and less reliable (Bandura, 1986, p. 44). A recent study by Pajares, Harlety and Valiante (2001) exploring the impact of different response formats in self-efficacy scales concluded that the 0-100 format was psychometrically stronger than a scale with a traditional Likert format.
- Items in a self-efficacy scale should be phrased in terms of "can do" instead of "will do". "Can is a judgment of capability, will is a statement of intention." (Bandura, 1986, p. 43)
- Response bias or social evaluation concerns should be minimized by using standard testing techniques such as assessing participants in private and without identification, and including an instruction that emphasizes frank judgment during self assessment and establishes the appropriate judgmental set; that is, participants are asked to judge their capabilities as of now, not their potential capabilities or their expected future capabilities (Bandura, 2001, p. 4).
- The hierarchical structure of self-efficacy surveys should either be random in order or ascending in order of task demands. Ordering items from most

to least difficulty can affect self-efficacy judgments (Bandura, 1986, p. 44).

- Measures of self-efficacy must be tailored to meet the specific "reading level" of the participants (Bandura, 2001, p. 4). Ambiguous or poorly worded items, technical jargon and multi-barreled items should all be avoided (Bandura, 2001, p. 4).

A preliminary 65-item scale was developed. Items were developed based upon discussion with "Internet experts" (i.e., academic and public librarians involved in teaching Internet use) and based upon the items used in the existing scales. The scale was piloted (n=5) and no concerns with "reading level" were noted. Before the scale can be administered, content validity will be determined by obtaining the feedback from both "Internet experts" and "self-efficacy experts". The feedback obtained will be used to make any necessary changes or refinements to the scale. Experts will be obtained from both the U.S. and Australia. Initial feedback was obtained from a poster outlining the scale and its construction at the 2003 Annual Convention for the American Psychological Society. The Internet self-efficacy scale created for use in the research is available on request from the author.

## Limitations of the Research

The research is limited by its reliance upon the use of self-reported measures by participants. Self reported measures provide a useful opportunity to collect data otherwise not readily available. But self reported data are limited by what "individuals know about their attitudes and are willing to relate" (Nunnally, 1967, p. 590). As such, a significant potential limitation in the current study is the overall validity of the measures employed. Despite the inherent problems, self-reported measures remain the most used and relied upon technique for data collection in the social sciences (Nunnally, 1967).

## Expected Outcomes and Significance of the Research

This research is significant because it develops a new theoretical framework through which to view the division between information "haves" and information "have-nots" within society. The research will illustrate that the digital divide involves more than just the availability of resources and funds to access those resources. It incorporates the internal forces of an individual that motivates him or her to use or integrate technology into his or her life. Using the Social Cognitive Theory by Bandura (1986) to examine these internal forces, this research will add another layer of understanding to the digital divide pyramid. The findings of the study will provide support to the existence of the Social Digital Divide as proposed by Harper (n.d.).

In addition, this is the first time that Internet self-efficacy has been explored within the context of the wider community. Existing studies that have examined self-efficacy have done so using university or high school students. The differences in these groups suggest that these studies cannot be generalised to the broader population. Equally important is that this is the first time a study exploring Internet self-efficacy and the digital divide will take place in Australia. The majority of studies to date have originated from the United States. The research will develop an Internet self-efficacy scale that is appropriate for use within the context of the general population.

This research is important because it expands current understanding of a phenomenon that has far reaching social and economic implications. The research will allow a more concise understanding of *what is* and *who represents* the digital inequality in society. Developing a clear and comprehensive picture of the forces behind the division in society between "haves" and "have-nots" is a vital step in bridging the gap. This research will allow organisations involved in the digital divide solution to develop and tailor services and programs to more accurately and effectively narrow the gap between information rich and information poor. As a consequence, real steps can be made in bridging the gap between the "haves" and the "have-nots" in society. It will allow for all members of community to have an equal chance of establishing and maintaining productive personal and professional lives in this rapidly emerging digital age.

# Future Research

In 2001 Luciano Floridi argued that "information and communication technologies have put humanity in charge of the world. We are the masters of the universe...The problem is that our ethical development has been much slower than our technological development" (Floridi, 2001b, p. 4). The research project outlined in this paper will help humanity to take steps to re-align our ethical and technological developments. More studies exploring the psychological, social and cultural barriers to the digital divide are needed. If we are to meet Floridi's challenge of developing an "information society for all" (2001a, p. 4) then we must more actively develop our understanding of the digital inequality within community. And this requires looking beyond the socio-economic explanation of digital equality. As suggested by Harper, "we need to re-conceptualise the [digital divide]…away from a simple lack of access and toward the social, cognitive, and communicative factors that truly divide groups" (2003, par. 2).

# Conclusion

The digital divide is a complex phenomenon. Developing a more sophisticated understanding of this phenomenon will aid organisations such as the public library in developing programmes and resources that can more effectively bridge the gap between information and technology "haves" and "have-nots". By examining the psychology of the digital divide using Bandura's Social Cognitive Theory (1986), this study will expand our current understanding of the digital divide and lend support to the existence of the Social Digital Divide as proposed by Harper (n.d.).

# Acknowledgment

The author would like to acknowledge the valuable comments and support provided by Associate Professor Christine Bruce, Dr. Gillian Hallam and Ms. Sylvia Edwards.

# References

Australian Bureau of Statistics (ABS). (2001). *Public libraries, Australia (8561.0)* [Online]. Available:*http://www.abs.gov.au/ausstats/abs@.nsf/ Lookup?NT0001AB66*

Bandura, A. (1977). Self efficacy towards a unifying theory of behavioural change. *Psychological Review, 84*(2), 191-215.

Bandura, A. (1986). *Social foundations of thought and action: A social cognitive theory.* Englewood Cliffs, NJ: Prentice Hall.

Bandura, A. (2001). *Guide for constructing self-efficacy scales.* Available on request from Professor Frank Pajares (mpajare@emory.edu).

Brisbane City Council. (2001). *Living in Brisbane 2010* [Online]. Available:*http://www.birsbane.qld.gov.au/council_at_work/planning/brisbane_2010/index.shtml*

Buchanan, E.A. (1999). An overview of information ethics issues in a world-wide context. *Ethics and Information Technology, 1,* 193-201.

Cassidy, S., & Eachus, P. (1998). *Developing the computer self-efficacy (CS) scale: Investigating the relationship between CSE, gender and experience with computers* [Online]. Available:*http://www.chssc.salfrd.ac.uk/healthSci/selfeff/selfeff.htm*

Churchill, G.A. (1992). *Marketing research.* Hinsdale, IL: The Dryden Press.

City of San Jose. (2001). Smart growth: Imagine a city.... *Inside San Jose* [Online], Fall/Winter. Available: *jose.ca.us/planning/sjplan/isjsmartgrowth.pdf*

Eastin, M.S., & LaRose, R. (2000). Internet self-efficacy and the psychology of the digital divide. *JCMS* [Online], *6*(1). Available:*http://www.ascusc.org/jcmc/vol6/issue1/eastin.html*

Floridi, L. (2001a). Information ethics: An environmental approach to the digital divide. *Philosophy in the Contemporary World* [Online], *9*(1), 39-46. Available: *http://www.wolfson.ox.ac.uk/~floridi/pdf/ieeadd.pdf*

Floridi, L. (2001b, May). *The new information and communication technologies for the development of education.* Invite Address. UNESCO World Commission on the Ethics of Scientific Knowledge and Technology (COMEST), Paris, [Online]. Available: *http://www.wolfson.ox.ac.uk/~floridi/pdf/eieu.pdf*

Foster, J.J. (2001). *Self-efficacy, African-American youth, and technology: Why the belief that "I cannot do that" is driving the digital divide.* Unpublished doctoral dissertation, University of Alabama, Tuscaloosa, Alabama, (UMI Dissertation Information Service, No. 3008531).

Foster, S.P. (2000). The digital divide: Some reflections. *International Information and Library Review, 23,* 437-451.

Hargittai, E. (2002). Second level digital divide: Differences in people's online skills. *First Monday* [Online], *7*(4), 1-7. Available: *http://firstmonday.org/issues/issue7_4/hargittai/index.html*

Harper, V.B. (n.d.). *Digital divide (DD): Redirecting the efforts of the scholarly community* [Online]. Available: *http://cal.csusb.edu/cmcrp/documents/Digital%20Divide%20position%20paper1(hypertext%20version).doc*

Harper, V.B. (2003) The digital divide (DD): A reconceptualization for educators. *Educational Technology Review* [Online], *11*(1). Available: *http://www.aace.org/pubs/etr/issue4/harper2.pdf*

Hnilo, L.A.R. (1997). *The hierarchy of self-efficacy and the development of an Internet self efficacy scale.* Available on request from the author.

Hongladarom, S. (n.d.). *Making information transparent as a means to close the global digital divide* [Online]. Available: *pioneer.netserv.chula.ac.th/~hsoraj/web/Making%20Info% 20Transparent.pdf*

Jung, J., Qiu, J.L., & Kim, Y. (2001). Internet connectedness and inequality: Beyond the digital divide. *Communication Research, 28*(4), 507-538.

Kvasny, L. (2002). *Problematizing the digital divide: Cultural and social reproduction in a community technology initiative.* Unpublished doctoral dissertation, Department of Computer Information Systems, Georgia State University, Georgia, USA [Online]. Available: *http://www.personal.psu.edu/faculty/l/ m/lmk12/KvasnyDissertation.pdf*

Lenhart, A., Horrigan, J., Ranie, L., Allen, K., Boyce, A., Madden, M., & O'Grady, E. (2003). The ever-shifting Internet population: A new look at Internet access and the digital divide. *The Pew Internet & American Life Project* [Online]. Available: *http://www.perinternet.org*

Maddux, J.E. (Ed.) (1995). *Self-efficacy, adaptation and adjustment: Theory, research and application.* New York: Plenum Press.

Maitland, C. (1996). *Measurement of computer/Internet self-efficacy: A preliminary analysis of computer self-efficacy and Internet self-efficacy measurement instruments.* Available on request from author.

Mosco, V. (1998). Myths along the information highway. In S. Berman & J. Danky (Eds.), *Alternative library literacy 1996 – 1997.* McFarland and Company, Jefferson.

National Office for Information Economy (NOIE). (2002). The current state of play [Online]. Available: http://www.noie.gov.au/projects/framework/Progress/ie_stats/CSOP_April2002/CSOP_April2002.pdf

National Telecommunication Information Association (NTIA). (1995). *Falling through the net: A survey of "have-nots" in rural and urban America* [Online].   Available: *http://www.ntia.doc.gov/ntiahome/fallingthru.html*

National Telecommunication Information Association (NTIA). (1999). *Falling through the net: Defining the digital divide* [Online]. Available: *http://www.ntia.doc.gov/ntiahome/fttn99/FTTN.pdf* .

National Telecommunication Information Association (NTIA). (2002). Falling through the net: Towards digital inclusion [Online]. Available: *http://www.ntia.doc.gov/ntiahome/dn/index.html*

Pajares, F., Hartley, J., & Valiante, G. (2001). Response format in writing self-efficacy assessment: Greater discrimination increases prediction. *Measurement and Evaluation in Counselling and Development, 33*(4), 214-221.

Sethi, V., & King, W.R. (1991). Construct measurement in information systems research: An illustration in strategic systems. *Decision Sciences, 22*(3), 455-472.

Stichler, R., & Hauptman, R. (Eds.). (1998). *Ethics, information, and technology*. McFarland, Jefferson.

Torkzadeh, G., & Van Dyke, T.P. (2001). Development and validation of an Internet self-efficacy scale. *Behaviour and Information Technology, 20*(4), 275-280.

## Chapter III

# Socio-economic Influence on Information Technology:
## The Case of California

Rasool Azari
University of Redlands, USA

James Pick
University of Redlands, USA

## Abstract

*This chapter examines the influence of socio-economic factors on the employment, payroll, and number of enterprises of three technology sectors for counties in California. Based on correlation and regression analyses, the results reveal that factors that are important correlates of technology sectors are professional/scientific/technical services, other services, and educational services, ethnicity, and college education. As a whole, the findings emphasize the importance of the association of socio-economic factors with the per capita magnitude of the technology sectors. This chapter suggests steps that can be taken by the state of California and its county and local governments to foster technology and reduce the digital divide.*

# Introduction

During recent years the rapid change in information technology (IT) and its impact on society have been the concern of academia, industry leaders, government officials and policy makers. There is no doubt that the impact of technology on society is profound and that it has long lasting effects financially, politically, and culturally. But the growing abundance of literature and projects concerning the social consequences of the IT and the Internet underscore the need for a better understanding of the forces at work.

Technological change has been central to the U.S. economic growth and is the major force in raising the nation's factor productivity at an accelerated rate. Information technologies (IT) are reshaping every aspect of organizations and business enterprises, such as work processes, decision-making, workforce, employment structures, teamwork, and products. "Indeed, the potential of the ICT revolution to transform the global economy has been at the centre stage in international forums and discussions..." (ILO, 2001, p. v). For companies to stay viable and competitive, adjusting to an ever-increasing pace of change is a must. The rapid development of new technologies in the information age and the unequal ability of societies across various segments to adjust to and assimilate these constant changes has been recognized as a source of problems for the old socio-economic structures because it creates potentially disruptive frictions.

This information gap is expressed by the term "digital divide," which is generally defined as "unequal access to information technology" (Light, 2001, p. 709). The effective utilization and accessibility of IT is the subject that some of the recent studies are trying to address. As Katz (2002, p. 4) puts it: "Having knowledge of what is there with no means of obtaining it or having technology but no knowledge of how to use it does not constitute access."

The continued existence of the "digital divide" and the increasing inequality of wages in the U.S. during the last two decades pose considerable challenges to policy makers. California, with its talented and diverse workforce, has a unique role in this equation. The long-term expansion of California's high tech, even with its recent slowdown, has and will depend on its skilled workforce (CCST, 2002). It has been recognized as the leading high-tech state in the U.S. (AEA, 2001). In the year 2000, among all states, it ranked first in high-tech employment, number of high-tech industry establishments, high-tech exports, R&D expenditures, and venture capital investment (AEA, 2001). It led the nation in all high-tech industry segments except photonics, and was second in high-tech per capita wage (AEA, 2001). Furthermore, 77 of its 1,000 private sector workers were employed by high-tech firms (AEA, 2001). Its R&D expenditure in 1998 was $43.9 billion, which was 19% of the nation's (AEA, 2001). California's leadership position in the technology sector justifies the importance of studies that analyze its sectoral growth in more detail and at smaller geographical scale, such as for counties in this chapter.

California slowed down in 2001; in particular the high-tech industry grew in California only by one percent, down sharply from 1999 and 2000, although this rate of growth varied from county to county (AEA, 2002). The slowdown is exacerbated by a large state government budgetary shortfall, which came to a head in 2002 and 2003, requiring stringent cutbacks and other actions. However, it is likely the state will recover from this problem in several years, and that it will not take away its technology leadership.

At the same time, there is a long-term weakness in its educational capability and readiness. "California is lagging behind other states in workforce readiness. Therefore its economic activities and slowdown, which include the Silicon Valley, much of the entertainment industry, and 48 federal government research labs, have repercussions on a global basis. If California cannot meet industry's demand for skilled labor, it could lose science and technology jobs to other states" (CCST, 2001; Conrad, 1999, p.1).

A recent study by CCST (2002) examined the reasons why California, even before the budget crisis, has fallen short in providing the requisite science and technology education to fulfill the high demand. This report points to substantial gaps in supply of skilled labor. For instance, the report points to a gap of 14,000 science and engineering works at the bachelor's degree level, out of 20,000 such degrees (CCST, 2001). Demographic growth has led to a huge increase in K-12 enrollments in the state, surging with a large proportion of immigrant students, while the education being provided is problematic (CCST, 2001). The report offers a flow diagram of production of high-tech workforce that centers on the public education system as well as workplace education and immigration to make up the deficits (CCST, 2001). The report finishes up with policy recommendations to increase the science and technology workforce that include strengthening educational programs at all levels, strengthening data-gathering and planning, and providing more marketing and funding for science and technology tracks (CCST, 2001). The report underscores a serious problem that in recent years has been solved by workforce immigration through special visas, although that is somewhat curtailed post 9/11.

Our study differs from other nationwide "divide" studies by emphasizing socio-economic correlates of information technology intensity for counties in California. It seeks to determine what social and economic factors are significant in distinguishing intensity of technology in three different industrial sectors in different counties in California. We consider the digital divide as a broad concept that includes economic, educational, and social aspects. For instance, a rich county economy is better able to afford technology and a highly educated county can better use it. Furthermore, social issues may also stimulate technology use. Since economic gains come, among others, from the adoption and adaptation of technology by individual users and by firms in business sectors, we use a framework that relates the socio-economic factors to the sizes of per capita employment, payroll, and enterprise prevalence in four different high-tech sectors in California. Among other things, we test whether or not education levels, community literacy in technology and presence of scientific/technical workforce are related to the size of county high-tech sectors. The aim here is to examine the associations between socio-economic factors and information technology for the counties in California. We are hoping that the results of this study will shed some light on the complex issues created by the emerging information technology by identifying some of the problems and thereby clarifying the ongoing dialogue among different stakeholders. This may help policy makers and experts to resolve California's already developing social and economic problems brought on because of the recent changes in IT and possibly avoid or prepare for potential others.

Following this introduction, we first review the literature on the digital divide and wage inequality and proceed with the research questions. We continue with the methodology used, the findings, discussion of the results, a section on policy implications and chapter limitations, and a conclusion.

# Background

The term "digital divide" entered the American vocabulary in the mid-1990s and refers to the unequal access to information technology (Light, 2001). There are slight differences among the various definitions of digital divide. Castells (2001) and Katz's (2002) definition of the digital divide refers to inequality of individual and household access to the Internet. OECD's definition (2000) is not limited to consumer (household) access to technology but also defines and distinguishes the level of penetration and diffusion of ICT in various sized enterprises.

The uneven distribution of IT benefits across the U.S is frequently pointed out. Major reports from the National Telecommunications and Information Administration, a unit of the Department of Commerce (NTIA, 1999, 2000, 2002) utilized the Current Population Survey of the U.S. Census to examine household distribution of access to technology including computers, phones, and the Internet. It defined "digital divide" as the divide between those with access to new technologies and those without. A key finding is that overall the nation "is moving toward full digital inclusion. The number of Americans utilizing electronic tools in every aspect of their lives is rapidly increasing" (NTIA, 2000, p. 4). By September 2001, 143 million Americans (ca. 54% of population) used the Internet and 174 million Americans used computers. The use of computers and the Internet is the highest among children and teenagers (NTIA, 2002). This rapid growth is among "most groups of Americans, regardless of income, education, race or ethnicity, location, age, or gender, suggesting that digital inclusion is a realizable goal" (NTIA, 2000, p. 4). Results of the NTIA data also show that ownership of a computer will increase the use of the Internet. So, access to computer is an important factor for the effective use of information technology and according to the NTIA data this ownership is growing rapidly.

Those Americans with high-income households enjoy much greater connectivity; other highly connected groups (holding income constant) are Asians, whites, and middle-aged, highly-educated, employed married couples with children. These are often found in urban areas and the West. On the other hand, low-income, Blacks, Hispanics, or Native Americans, seniors in age, not employed, single-parent (especially female) households, those with little education, and those residing in central cities or especially rural areas are the least connected (NTIA, 1999).

However, "a digital divide still remains" (NTIA, 2000, p. 15). Even though the utilization of computers expanded dramatically over a 15-year period for all groups in the U.S., a digital divide persisted and in some cases grew (NTIA as discussed in Noll et al., 2001). For instance in 1989, the percent of U.S households that owned a computer, by educational level—elementary education and bachelor's or more education—was 1.9% and 30.6% respectively. In the year 2000, these numbers have changed to 12.8% for elementary and 75.7% for bachelor's or more education. The Hispanic-white ethnicity differences widened from 7.1 versus 16.0% in 1989 to 33.7 versus 55.7% in 2000 (NTIS data cited in Noll et al., 2001).

Furthermore, Lentz (2000) argues that the concept of the digital divide should be used in broader terms than merely describing end user problems and should extend also to community development. Other researchers apply the term to business, economy, and/or society levels, rather than the individual level (Baker, 2001; OECD, 2000). Baker points out that the policy problem of the digital divide is best addressed through multiple dimensions, that is, policies that address disparities in information technology diffusion at different geographic, economic, social, and organizational levels. In their book, *Social Consequences of Internet Use,* Katz and Rice (2002) analyzed the impact of the Internet on American society from three different perspectives—access, involvement, and interaction. "The book's three sections correspond to these three vital issues concerning Internet and human communication—access, civic, and community involvement, and social interaction and expression" (Katz, 2002, p. 4).

Florida and Gates (2001) examined the correlations of diversity indices with a high-tech index in the 50 largest metropolitan areas in the U.S. in 1990. The authors conclude that cultural, ethnic, and social diversity are linked to high tech. They do not prove causality, but speculate about the reasons for this. Among the explanations are that high-tech thrives in a creative atmosphere, which comes from community members (i.e., human capital) who are not mainstream or traditional but differ in their perspective, and are often new arrivals. Governments in metropolitan areas can foster community atmospheres that attract such diversity. College education was also shown to be linked to high tech as strongly as the composite diversity index (Florida & Gates, 2001).

Another issue that widens this digital gap is the phenomenon of wage divergence and inequality related to technological change. This issue has received the attention of many labor and trade economists (Feenstra, 1997) and it is widely believed that the development of the new technology increases the demand for skilled workers, thereby increasing the wage differential between skilled and unskilled workers. Even though empirical evidence from the literature on wage inequality is inconclusive and fragmented (Deardorff, 1998, p. 371), there is still a general consensus among many economists that technological change is the primary explanation for the widening gap in inequality of wages in the United States.

In addition, geography plays an important part in studies such as this one. Nations, states, counties, and cities have diverse features and characteristics that leave their particular marks. At the city level in the U.S., a wealth of high-tech data appeared in *California Cybercities,* published by the American

Electronic Association (AEA, 2002). This study supports the large extent of the digital divide. Presenting data rather than analyzing them, this volume is useful as a data source, some of which is included in the present paper. It provides an overview of California's high-technology industry in the eight major metropolitan areas—Los Angeles, Oakland, Orange County, Sacramento, San Diego, San Francisco, San Jose, and Ventura.

The economy of California and its large technology sector started slowing in 2001. Many high-tech companies announced layoffs in 2001 and 2002 (AEA, 2002). San Jose, the leading high-tech city in California, experienced a dramatic decline in technology employment in 2001; of the eight metropolitan areas examined in 2002, Los Angeles, Oakland, and Sacramento had some high-tech employment growth in 2001 but a much lower growth rate than in 2000; the slowdown in the San Francisco and San Jose area was particularly high and many companies from there moved to the Oakland and Sacramento areas because of increases in business costs, particularly for land and labor. In 2002, San Jose ranked first in the concentration of high-tech workers, San Francisco second, followed by Oakland (AEA, 2002).

From the industry sector employment perspective the AEA reports that San Jose led in nine out of 13 technology industry sectors in 2000; San Diego led in consumer electronics manufacturing, Los Angeles led in defense electronics manufacturing, communication services, and data processing and information services; Orange County ranked second in six of the 13 electronic sectors. These technology sectoral differences stem from particular and often complex aspects of the development of California's metro areas in the mid to late 20th century (see Kling, Olin and Poster, 1991, for an explanation of such phenomena in Orange County).

San Jose ranked first and San Francisco second in the compensation in the high-tech industry. Even in the slowed economy of 2001, the high-tech workforce for this industry was still paid significantly higher than average private sector workers; for example, in San Diego the earning of high-tech workers was 132% more than for the private sector workers, and in Sacramento there was a 126% differential. A point of interest here is the wage inequality. This supports the earlier contention of a widening wage inequality in the U.S.

Nationwide studies of socio-economic correlates of the sizes of technology sectors at the county level point to the importance of highly skilled professional/ scientific/technical labor force (PST), ethnicity, and college education, and to a lesser extent support services, income, and federal funds grants (Azari &

Pick, 2003a, 2003b). The primary finding of the association of PST with technology emphasizes the key factor of highly skilled workforce both to perform technology tasks and provide highly skilled support functions. The secondary finding of positive links to particular ethnicities supports the point that community diversity is positively tied to technology. These studies utilized methods and data somewhat similar to the present study. However, they differ by not separating out the nation's highest tech state of California and considering its particular technology, economic, and social attributes, and how that informs its policy options.

The many findings emerging from these and comparable studies support that socio-economic factors impact the use of IT. Furthermore, they suggest the causality depends on the unit of analysis, region, and choice of dependent variable, such as IS utilization impact, application area, or IT investment. As stated before, we consider the digital divide as a broad concept that includes economic, educational, and social aspects. For instance, a rich economy is better able to afford technology and a highly educated community can better utilize technology. Furthermore, social issues may also stimulate technology use. For example, socio-economic characteristics influence consumer uses of technology. In turn, consumers with scientific and technology skills provide technology employees for businesses. Those employees contribute to corporate receipts and payrolls. Cumulative corporate results constitute technology sectors in counties.

We believe that socio-economic factors and technological development are interrelated. "Technology is deeply affected by the context in which it is developed and used. Every stage in the generation and implementation of new technology involves a set of choices between different options. A range of interrelated factors—economic, social, cultural, political, organizational— affect which options are selected. Thus, a field of research has emerged where scholars from a wide variety of backgrounds seek to increase the understanding of technological development as a social process" (Dierkes & Hoffman, 1992). Today technology has become so intertwined with our everyday life that a broad understanding of its utilization and distribution requires a thorough examination of the socio-economic environment. Paying attention to the relationship between socio-economic factors and the changes in the high-tech sectors may further shed some light on the problem and help to alleviate the digital divide.

# Research Questions

This chapter has two research questions:

1. What are the most important socio-economic factors overall that influence the per capita economic sizes of the information, information services/data processing, telecommunications/broadcasting, and motion picture/sound recording technology sectors for counties in California?

2. How do these sectors differ with respect to the most important socio-economic factors that influence their economic sizes?

# Methodology

This chapter investigates the association of technological development with socio-economic factors for counties in California. The data, all for year 2000, were from the U.S. Census of 2000, City and County Databook, and County Business Patterns detailed North American Industry Classification System (NAICS) data (U.S. Census, 2001, 2003a, 2003b). For each regression or correlation sample, all of California's 58 counties were initially included. However, in the correlation and regression samples, counties with missing data were excluded.

The basic assumption is that socio-economic factors, such as education, income, service sector composition, ethnicity, federal funding, and population growth, are associated with the size of technology receipts and payroll. It is evident that socio-economic factors and technology are frequently intertwined and interrelated. For instance, in a county, the presence of a wealthy population, a highly educated labor force, colleges/universities, and high R&D expenditures will likely increase the technology level by attracting capital investment and stimulating productive enterprises. At the same time, a county with a high technology level may attract highly educated people; may create prosperity and wealth in its citizenry; and may foster R&D.

Our research framework is based on the unidirectional relationship between the socio-economic factors and technological change as depicted in the figure. These factors are one of the most important variables in social and economic

*Figure 1.*

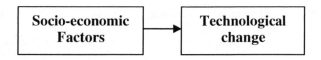

development studies. A study of bi-directional effects would require more sophisticated data collection and intermediate variables than are undertaken in this study.

It is clear that this framework may also flow in a feedback way in the opposite direction, in that larger county technology sectors may attract populations with certain socio-economic characteristics. We are not ready to present a comprehensive framework of these different dimensions and their linkages. Rather, this chapter focuses on unidirectional linkage as shown in Figure 1.

The socio-economic factors for this research are:

- professional/scientific/technical workforce,
- educational services workforce,
- other services workforce,
- median household income,
- college graduates,
- federal funds and grants,
- change in population 1990-2000, and
- proportions of Black, Asian, Latino, white, and female.

The unit of analysis is the county, because it is the smallest geographical unit for which a wide range of statistical data can be obtained. Its unit applies across the entire state, has accurate and extensive variables collected through the U.S. Census and other sources, and is stable geographically over time. It also represents a governmental and policy unit, so policy suggestions from research can inform governmental decision-making.

The selection of independent variables is based on prior studies. For example, counties with higher service components will also have higher levels of technology. We justify this variable by the substantial proportion of high-tech employees who work in services versus other sectors, and the growth in this proportion over time. For instance, in 2000, 63% of high-tech workers were in services, a proportion 11% higher than in 1994 (AEA, 2001). Hence, we reason that more service-oriented counties will have higher levels of technology workers, and the technology sectors will be more ample. We also reason that technology sectors will have lower levels in counties that receive higher levels overall of federal funding on a per capita basis. The federal grants include a large proportion of grants for social services, welfare, and community improvements. On a per capita basis, we reason that federal grants would be more prominent in poorer counties.

Differences in technology utilization are also based on demographic attributes as highlighted by the U.S. Department of Commerce studies (NTIA, 1999, 2000, 2002). For instance, in 1998 the proportion of the population using the Internet varied between 8% for very low income households (under $20,000) to 60% for high income ones (over $75,000) (NTIA, 1999). At the level of developing nations, per capita income was one of the most important influences on Internet and cell phone use (Dasgupta et al., 2001). Likewise, 47% of white households had a computer, versus 23 and 25% respectively for Blacks and Hispanics. In a recent publication, Lentz (2000) emphasizes that race and ethnicity should not be excluded from studies of high-tech jobs. Slowinski (2000) points to large differences between the ethnic groups as a cause of differences in the high-tech workforce. Crandall (2001) identifies ethnicity and income as important variables for access to computers and technology. Florida and Gates (2001) and Florida (2002) demonstrate a positive association between diversity and technology. They also show a strong relationship between college education and technology.

The U.S. Department of Commerce identified a gender gap, with about 3% more utilization of the Internet by surveyed males versus females (NTIA, 1999). We included change in county population based on our own reasoning that growing counties draw in workforce and capital investment, and tend to have growing educational institutions that encourage technology use. This attribute was included in two prior studies (Azari & Pick, 2003a, 2003b), although it was shown to be of minor significance.

Because socio-economic variables have proven to be of importance, we feel justified in including them, even though the units of analysis (i.e., individuals, firms, and nations) have varied in much of the prior research. We reiterate that although the directionality of effect may be in two directions, we emphasize in this study how socio-economic attributes influence technology level. Multicollinearity (Neter, 1996) is a minor issue because of the use of stepwise regression technique with mostly only one or two independent variables entered.

To measure technological change, the dependent variables are employees per capita, payroll per capita, and number of enterprises per capita. The information sectors selected are:

- Information (broad sector),
- Information Services/Data Processing (IS/DP) — subsector of information,
- Broadcasting/Telecomm (B/T) — subsector of Information, and
- Motion Picture/Sound Recording (MP/SR) — subsector of Information.

For the Information Sector, only number of enterprises per capita is included, due to very high inter-correlations with the other dependent variables.

The following are several reasons why we selected these information sectors:

1) they utilize computers and modern information communication technologies;

2) they increase the productivity of institutions, shorten product life cycle, and reverse the composition of our labor force from mainly blue collar workers to predominantly service providers and knowledge workers;

3) they diminish the importance of distance and contribute to globalized markets and economies; and

4) they contribute nearly 60% to the American gross national product.

In summary, we explore the association of 12 independent socio-economic factors with three dependent variables, for four industry sectors. Correlation and stepwise linear regression analyses were conducted to test our models. The software utilized was SPSS version 11.

# Findings

The dependent variables are inter-correlated with the exception that MP-SR variables are not associated with those for IS-DP and B-T (see Table 1). The entertainment industry (movies, recording) does not correspond to the geographic patterns of information services and communications. This is not surprising on a common sense basis, since for instance computers and information technologies are centered in the Silicon Valley, while entertainment is centered in greater LA.

The regression findings for technology employment and payroll are all significant and demonstrate that the most important factor is PST employment, followed by education (see Table 2). PST is the most important in estimating IS-DP and B-T employment and payroll. The secondary factors are educational services (in one regression each for IS-DP and B-T) and percent Black for B-T payroll. On the other hand MP-SR is correlated with educational services and college graduates, as well as with percent change in population for MP-SR payroll.

Regressions for the number of technology enterprises are likewise all significant. PST and education are again the most significant, but education assumes a greater role than for employment and payroll. As seen in Table 3, PST is the most important predictive factor for the information sector as a whole, and for IS-DP, but it is not significant for B-T and MP-SR. Education is important for each regression — college graduates for the information, IS-DP, and B-T sectors, and educational services for the MP-SR sector. Federal grants are positively significant for IS-DP and B-T. Finally two ethnic variables, percent Asian for information sector and percent Latino for IS-DP, significantly reduce the number of enterprises.

# Discussion

The dominance of PST as a predictive factor of California's county technology sectors, as seen in Tables 1 and 2, corresponds to other nationwide county studies (Azari & Pick, 2003a, 2003b). In the case of California, it is the most prominent state in its PST human resource (AEA, 2001). However, the state's PST workforce per capita is varied in its distribution, and corresponds to

*Table 1. Correlation matrix of dependent variables, California sample, 2000*

| | No. enterprises/capita INF | No. employees/capita IS-DP | Payrolls/capita IS-DP | No. enterprises/capita IS-DP | No. employees/capita B-T | Payrolls/capita B-T | No. enterprises/capita B-T | No. employees/capita MP-SR | Payrolls/capita MP-SR |
|---|---|---|---|---|---|---|---|---|---|
| No. enterprises/capita INF | 1.000 | | | | | | | | |
| No. employees/capita IS-DP | 0.679*** (n=31) | 1.000 | | | | | | | |
| Payrolls/capita IS-DP | 0.640*** (n=31) | 0.859*** (n=31) | 1.000 | | | | | | |
| No. enterprises/capita IS-DP | 0.898*** (n=47) | 0.814*** (n=31) | 0.798*** (n=31) | 1.000 | | | | | |
| No. employees/capita B-T | 0.678*** (n=44) | 0.8575** (n=31) | 0.543** (n=31) | 0.731*** (n=42) | 1.000 | | | | |
| Payrolls/capita B-T | 0.713*** (n=44) | 0.612*** (n=31) | 0.582*** (n=31) | 0.791*** (n=42) | 0.960*** (n=44) | 1.000 | | | |
| No. enterprises/capita B-T | 0.617*** (n=57) | 0.544** (n=31) | 0.511** (n=31) | 0.613*** (n=47) | 0.754*** (n=44) | 0.737*** (n=44) | 1.000 | | |
| No. employees/capita MP-SR | 0.732*** (n=30) | 0.283 (n=30) | 0.177 (n=30) | 0.478** (n=30) | 0.243 (n=30) | 0.266 (n=30) | 0.273 (n=30) | 1.000 | |
| Payrolls/capita MP-SR | 0.617*** (n=30) | 0.150 (n=30) | 0.067 (n=30) | 0.350 (n=30) | 0.108 (n=30) | 0.134 (n=30) | 0.161 (n=30) | 0.967*** (n=30) | 1.000 |
| No. enterprises/capita MP-SR | 0.719*** (n=54) | 0.321 (n=31) | 0.179 (n=31) | 0.464*** (n=46) | 0.319* (n=43) | 0.338* (n=43) | 0.248 (n=54) | 0.929*** (n=30) | 0.829*** (n=30) |

Note: the top of each cell gives the Pearson correlation and the bottom sample size

Note: INF = Information Sector
IS-DP = Information Services-Data Processing Subsector
B-T = Broadcasting-Telecommunications Subsector
MP-SR = Motion Picture, Sound Recording Subsector

\* correlation significant at 0.05 level
\*\* correlation significant at 0.01 level
\*\*\* correlation significant at 0.001 level

*Table 2. Standardized regression results for dependent variables, no. of employees and payroll, per capita, California sample, 2000*

| | Employees per Capita for IS-DP Beta Value | signif. | Payroll per Capita for IS-DP Beta Value | signif. | Employees per Capita for B-T Beta Value | signif. | Payroll per Capita for B-T Beta Value | signif. | Employees per Capita for MP-SR Beta Value | signif. | Payroll per Capita for MP-SR Beta Value | signif. |
|---|---|---|---|---|---|---|---|---|---|---|---|---|
| Professional/ Scientific/ Technical Employees per capita | 0.824 | 0.000*** | 0.457 | 0.020* | 0.665 | 0.000*** | 0.763 | 0.000*** | | | | |
| Educational Services Employees per capita | | | 0.446 | 0.023* | 0.252 | 0.027* | | | 0.552 | 0.002** | | |
| Other-Services Employees per capita | | | | | | | | | | | | |
| Median Household Income (in dollars) | | | | | | | | | | | | |
| Federal Grants and Funds (in 1000's of dolllars) | | | | | | | | | | | | |
| College graduates per capita | | | | | | | | | | | 0.531 | 0.003** |
| Percent Change in Population 1990-2000 | | | | | | | | | | | 0.334 | 0.046* |
| Proportion of Population 65+ | | | | | | | | | | | | |
| Percent Black | | | | | | | 0.216 | 0.029* | | | | |
| Percent Asian | | | | | | | | | | | | |
| Percent Latino | | | | | | | | | | | | |
| Percent Female | | | | | | | | | | | | |
| Regression adjusted R sqaure | 0.669 | | 0.704 | | 0.601 | | 0.667 | | 0.280 | | 0.329 | |
| significance level | | 0.000*** | | 0.000*** | | 0.000*** | | 0.000*** | | 0.002** | | 0.005** |
| sample size (N) | 31 | | 31 | | 31 | | 31 | | 31 | | 31 | |

\* signif. at 0.05
\*\* signif. at 0.01
\*\*\* signif at 0.001

Note: INF = Information Sector
IS-DP = Information Systems-Data Processing Subsector
B-T = Broadcasting-Telecommunications Subsector
MP-SR = Motion Picture, Sound Recording Subsector

county technology sectors. At one extreme, San Jose County has one of the greatest concentrations of PST workers in the nation, since it includes Silicon Valley's R&D establishment, famous scientific universities such as Stanford and U.C. Berkeley, and advanced government research centers such as NASA-Ames. The environment of discovery and science also is associated with a high concentration of technology sectors in computers, information systems, communications, the Internet, and Web entertainment. On the other hand, an opposite example is Stanislaus County in the far north, which has among the lowest levels in the sample of both PST per capita and technology

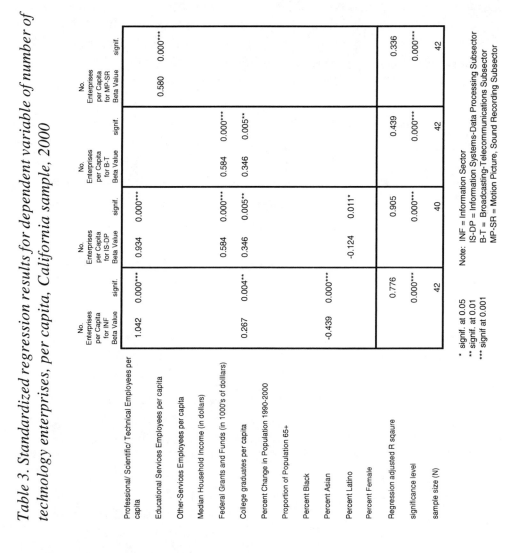

*Table 3. Standardized regression results for dependent variable of number of technology enterprises, per capita, California sample, 2000*

|  | No. Enterprises per Capita for INF Beta Value | signif. | No. Enterprises per Capita for IS-DP Beta Value | signif. | No. Enterprises per Capita for B-T Beta Value | signif. | No. Enterprises per Capita for MP-SR Beta Value | signif. |
|---|---|---|---|---|---|---|---|---|
| Professional/ Scientific/ Technical Employees per capita | 1.042 | 0.000*** | 0.934 | 0.000*** |  |  |  |  |
| Educational Services Employees per capita |  |  |  |  |  |  | 0.580 | 0.000*** |
| Other-Services Employees per capita |  |  |  |  |  |  |  |  |
| Median Household Income (in dollars) |  |  |  |  |  |  |  |  |
| Federal Grants and Funds (in 1000's of dolllars) |  |  | 0.584 | 0.000*** | 0.584 | 0.000*** |  |  |
| College graduates per capita | 0.267 | 0.004** | 0.346 | 0.005** | 0.346 | 0.005** |  |  |
| Percent Change in Population 1990-2000 |  |  |  |  |  |  |  |  |
| Proportion of Population 65+ |  |  |  |  |  |  |  |  |
| Percent Black |  |  |  |  |  |  |  |  |
| Percent Asian | -0.439 | 0.000*** |  |  |  |  |  |  |
| Percent Latino |  |  | -0.124 | 0.011* |  |  |  |  |
| Percent Female |  |  |  |  |  |  |  |  |
| Regression adjusted R sqaure | 0.776 |  | 0.905 |  | 0.439 |  | 0.336 |  |
| significance level | 0.000*** |  | 0.000*** |  | 0.000*** |  | 0.000*** |  |
| sample size (N) | 42 |  | 40 |  | 42 |  | 42 |  |

* signif. at 0.05
** signif. at 0.01
*** signif at 0.001

Note: INF = Information Sector
IS-DP = Information Systems-Data Processing Subsector
B-T = Broadcasting-Telecommunications Subsector
MP-SR = Motion Picture, Sound Recording Subsector

sector sizes per capita. The full extent of this range, extending to rural, agricultural California, needs to be recognized.

The secondary yet very important educational factors are also key to fostering technology. These factors influenced the number of technology enterprises for the information, IS-DP, and B-T sectors (see Table 1) and affected payroll per capita for MP-SR (Table 2). CCST and other reports have pointed to a severe shortage of science and technology graduates, relative to the high demand of the nation's leading technology state (CCST, 2002). CCST ascribes this shortage to a number of structural weaknesses in K-12 and college education. Not enough science and technology opportunities are present in K-12, and there is reduced incentive and resultant low numbers of science and engineering majors and the college level (CCST, 2002). Since California cannot produce sufficient skilled workers, and it is not currently receiving large domestic migration, the gap is being filled by highly skilled workforce immigrating from overseas, many of whom are being admitted to the U.S. on temporary visas for special needs (CCST, 2002). Getting back to our finding of the importance of education, this points to an increasingly dysfunctional state educational system and the need for policy changes that are discussed later.

The lack of PST affects MP-SR seen for both number of technology enterprises (Table 1), and employees per capita and payroll per capita (Table 2). It may be ascribed to lack of connection between a general scientific/R&D climate and growth in the movie/recording industries. They may derive less benefit from basic science, engineering, or R&D, since they are not as dependent on a science climate. For instance, many aspects of movie making are not taught widely in universities, studied in government labs, or dependent on scientists. On the other hand, movies and recording industries do benefit by good education and abundant college graduates. The correlation analysis in Table 1 had detected that the values for MP-SR's independent variables were predominantly unrelated to those for IS-DP and B-T, so these different regression findings are not surprising.

Among the lesser factors in the findings, ethnicity effects only appear for three out of ten regressions, more strongly as an inverse effect of percent Asian on the number of technology enterprises for the information sector (see Table 3), and more weakly for the Latino and Black percentages in two other cases, seen in Table 2. They were also opposite from the hypothesized positive direction in two of three cases. The strong Asian effect is surprising and unexplained, given the major involvement of Asians in many technology fields in California. Further research is needed to determine the robustness of this finding and to determine its cause. Ethnicity's two weaker findings and absence in the remaining regressions do not lend backing, for California, to a recent framework linking diversity to technology (Florida, 2002; Florida & Gates, 2001). The explanation may be that this is due to the high level of diversity in the entire state. According to the U.S. Census (2003), its non-white percentage in 2000 was 40.5%. This means that diversity is reduced in prominence and overwhelmed by PST and education. Translating this into an everyday explanation, it implies that for a highly diverse state, there is an overall creative contribution stemming from the diversity, but that the *differences* in technology levels between counties are much more due to PST and education. The single positive association between Black population and payroll per capita for B-T is consistent with the diversity theories and several prior studies. We do not have an explanation for this specific effect, especially since it applies to B-T payroll and not to B-T employment.

There are similarities and differences between the California and nationwide county findings with somewhat similar regression design, variables, and data (Azari & Pick, 2003a, 2003b). The current study's dominant importance for PST corresponds to the national findings. However, other positive factors appear for the nation, namely other services and household income, while ethnicity factors were more prevalent but inverse. At the same time, education was less prominent nationally (Azari & Pick, 2003a, 2003b). It would appear that California's educational deficit for technology workers pushes this factor to greater importance, while its high base level of diversity reduces that factor's importance. Those far-reaching ideas can be analyzed with more detailed future studies that, for instance, include data on state or county science/technology educational deficits and on perceptions of diversity.

It is worth mentioning several limitations of this study. Due to the geographical proximity of counties and the problem of urban sprawl in California, especially in metropolitan areas, there is a likelihood of daily workforce inter-county commuting. This limitation can be addressed in a future study that includes commuting flows and characteristics of commuters. Another limitation is the application of the stepwise multiple regression method. Regression results are estimates of the changes that would occur if the variables were entirely independent of one another. In complex social phenomena, such as those addressed in this study, single factor changes are rare. For example, income, education, and occupation are unlikely to change separately without inducing changes in the other components. Such cross-dependencies highlight the dangers of interpreting results as estimates of causal changes.

A further study limitation is that there may be extraneous factors that affect the analysis, such as laws and regulations; profit-making versus societal interests of technology firms; high-tech employment trends; venture capital investment; public and private technology education; and cultural, sociological and behavioral factors that influence businesses and their workers. Many are not available consistently for counties from the U.S. Census, but may be from other sources. Also, the study is limited by the model's unidirectionality, which goes from socio-economic factors to technology sector sizes. An expanded study could be based on a bidirectional feedback model, mediated by intermediate factors.

# Policy Implications and Future Research

The results of this study have various implications for different stakeholders—individuals, organizations, the government, and society as a whole. The following section addresses some of these concerns.

Since the debate on the digital divide ranges across individual consumers, businesses, technology sectors, counties and states, and whole economies, it is appropriate to indicate that these levels are inter-connected. As noted before, socio-economic characteristics influence consumer uses of technology. In turn, consumers with scientific and technology skills provide technology employees for businesses. Those employees contribute to corporate receipts and payrolls. Corporate results add up to constitute technology sectors in counties. This framework may also flow in a feedback way in the opposite direction, in that larger county technology sectors may attract populations with certain socio-economic characteristics. Changes in technology demand continuous skill upgrading in the work force. Even though much remains to be done, the computer and ICT intensive industries are showing the greatest increase in the skill upgrading of their work force in California. This trend, though, bypasses many talented individuals and leaves a major part of the population behind by limiting their chance to participate in and gain from the economic growth.

The importance and advantages for technology development of scientists, professionals, and college-educated workforce are stressed in this chapter. A college educated environment leads to improved skills and the ability to educate and retrain broad parts of the community. All this fosters technology industries and enterprise. This impact may be limited to technologically-skilled employees and hence the potential gains associated with this transformation are likely to be limited, which in turn means the digital divide is deepening. This dynamic needs to be adjusted by appropriate institutions and policy choices. The extent of the success of technological change depends also very much on the socio-economic context. Policy intervention must be targeted towards reducing the gap in social and economic factors since this study has shown that a direct influence exists between specific socio-economic factors and the dependent variables.

Preventing the divide from increasing, though, is no easy task. A big problem with technology arises from its inadequate distribution to society, and consequently, its unequal access to, and use by some members of society. But focusing solely on the more equal distribution of technologies will not solve the problem of the digital divide. Baker (2001) points to the additional influences of education, geography, race and income as indicated by studies conducted by the U.S. Department of Commerce (NTIA, 1999, 2000, 2002) and other studies (Benton, 1998; CCST, 2001; Hoffman, 1998). The technology gap existing in the U.S. was the subject of investigation by the Benton Foundation (1998), which arrived at "the conviction that the design of the communication system through which we will talk to one another, learn from one another, and participate in political and economic life together is too important to be left to the free market alone" (Lentz, 2000, p. 362).

California historically has been the leader of many social and political movements. Its policy implementations have repercussion worldwide. Being also a leader in high-tech sectors, California has the potential for influencing the problem of the digital divide and wage inequality. Part of the planning focus in California for technology needs to be on improving the general social, educational, and economic levels of counties.

Overall, the findings of this paper reinforce the notion that the digital divide exists and its causes are complex. To alleviate this problem the following policies are suggested for California and its counties:

- Promote and support science and high-end professions by investing in these segments of higher education and in R&D.
- Invest in broadly-based education, training, and lifelong learning.

- Develop policies to market science and high-end professional career paths to K-12 and college students.

- Create educational incentives for students to enter those career paths.

- Shift the dependence on reducing the high-tech educational gap from foreign immigration to the state's education system and its outflows.

- Shift the focus of unemployment protection to skills upgrading and lifelong education.

- Increase awareness of the negative and exclusionary effects the digital divide has on individuals and society in general and its possible subsequent results—unemployment, social unrest, and crime.

- Establish a comprehensive dialogue between different stakeholders in communities—government, businesses, educational, institutions, citizens— to foster a unified approach to encourage science and educational development and its benefit to all.

- Allocate resources more efficiently.

The current research project can be expanded in a number of ways. Other high-tech states and regions can be studied and compared, such as Washington D.C. and Boston. A multi-nation study of socio-economic factors and technology levels with counties or states as units of analysis could be conducted in order to establish if there are cultural and international differences. That study should include similar variables such as age, gender, ethnicity, education, high-tech employment trends, venture capital investments, international trade, high-tech average wage, R&D per capita, and technology transfer.

# Conclusion

This chapter's focus is on two research questions. In conclusion, the study's statistical results, which have been described and discussed, support both of the questions and provide the following answers to them.

1.  *What are the most important socio-economic factors overall that influence the per capita economic sizes of the information, informa-*

*tion services/data processing, telecommunications/broadcasting, and motion picture/sound recording technology sectors for counties in California?*

The most important factor is professional/scientific/technical workforce. The next factors in importance are college graduates and educational services. Of minor importance are ethnicity (with inconsistent directionality), federal grants and funds, and the very minor percent change in population.

2. *How do these sectors differ with respect to the most important socio-economic factors that influence their economic sizes?*

By sectors, the information sector is influenced by PST and education about equally. That sector shows some positive effect from federal funds and grants, and minor inverse ethnic effects. IS-DP and B-T are similar in having a dominant PST effect, followed by lesser educational effects. There are specific, minor Latino and Black ethnicity effects. For MP-SR, educational services and college graduates dominate, with a minor percent population change effect.

These conclusions suggest that to develop its technology sectors further, California counties need to plan for and invest in building up more their scientific and professional workforce and related educational and scientific institutions, and to encourage educational opportunities and services. In a time of governmental budget constraints, California county governments and community and business leadership groups should strive to not lower their support for these factors, in order to maintain the counties' technology edge.

# References

American Electronics Association. (2001). *Cyberstates 2001: A state-by-state overview of the high-technology industry.* Santa Clara, CA: American Electronics Association.

American Electronics Association. (2002). *California Cybercities 2002.* Santa Clara, CA: American Electronics Association.

Azari, R., & Pick, J.B. (2003a). The influence of socio-economic factors on technological change: The case of high-tech states in the U.S. In R. Azari (Ed.), *Current security management and ethical issues of information technology* (pp. 187-213). Hershey, PA: IRM Press.

Azari, R., & Pick, J.B. (2003b). Technology and society: Socio-economic influences on technological sectors for United States Counties. *International Journal of Information Management,* in press.

Baker, P.M.A. (2001). Policy bridges for the digital divide: Assessing the landscape and gauging the dimensions. *First Monday* [Online], *6*(5). Available: *www.firstmonday.org*

Benton Foundation. (1998). *Losing ground bit by bit: Low-income communities in the information age.* Report. Washington, DC: Benton Foundation.

CCST. (2001). *Bridging the digital divide.* Sacramento, CA: California Council on Science and Technology.

CCST. (2002). *Critical path analysis of California's science and technology education system.* Report. Sacramento, CA: California Council on Science and Technology.

Conrad, A.C. (1999). *Industry sector analysis of the supply and demand of skilled labor in California.* Report. Sacramento, CA: California Council on Science and Technology.

Crandall, R.W. (2001). The digital divide: Bridging the divide naturally. *Brookings Review, 19*(1), 38-41.

Dasgupta, S., Lall, S., & Wheeler, D. (2001). *Policy reform, economic growth, and the digital divide: An econometric analysis.* Research Paper, World Bank. Washington, D.C.: Development Research Group.

Deardorff, A.V. (1998). Technology, trade, and increasing inequality: Does the cause matter for the cure? *Journal of International Economic Law,* 353-376. Oxford University Press.

Dierkes, M. (1992). 21st century technologies: Promises and perils of a dynamic future. *OECD.*

Feenstra, R.C., & Hanson, G.H. (1997, June). *Productivity measurement, outsourcing, and its impact on wages: Estimates for the US, 1972-1990.* NBER Working Paper, No. 6052.

Florida, R. (2002). *The rise of the creative class.* New York: Basic Books.

Florida, R., & Gates, G. (2001). *Technology and tolerance: The importance of diversity to high-technology growth.* Washington, D.C.: The Brookings Institution.

Graham, S. (2002). Bridging urban digital divides? Urban polarisation and information and communications technologies (ICTs). *Urban Studies, 39*(1), 33-36.

Hoffman, D.T.N., & Venkatesh, V. (1998). Diversity on the Internet: The relationship of race to access in usage. In A. Garmer (Ed.), *Investing in diversity: Advancing opportunities for minorities and the media* (p. 130). Washington, D.C.: The Aspen Institute.

International Labor Office. (2001). *World employment report: Life at work in the information economy.* Geneva: ILO.

Katz, E. J., & Rice, R.E. (2002). *Social consequences of Internet use: Access, involvement, and interaction.* Cambridge, MA: MIT Press.

Kling, R., Olin, S., & Poster, M. (1991). *Postsuburban California: The transformation of Orange County since World War II.* Berkeley: University of California Press.

Lentz, R.G. (2000, August). The e-volution of the digital divide in the U.S.: A mayhem of competing metrics. *Info, 2*(4), 355-377.

Light, J. (2001). Rethinking the digital divide. *Harvard Educational Review, 71*(4), 709-733.

Miller, D.C. (1991). *Handbook of research design and social measurement* (5th ed.). Sage Publication.

National Telecommunication Information Administration (NTIA). (2000). *Falling through the Net: Towards digital inclusion.* Washington, D.C.: U.S. Department of Commerce.

National Telecommunication Information Administration (NTIA). (2002). *A nation on line: How Americans are expanding the use of Internet.* Washington, D.C.: U.S. Department of Commerce.

Neter, J., Kutner, M.H., Nachtsheim, C.J., & Wasserman, W. (1996). *Applied linear statistical models.* Boston: WCB/McGraw-Hill.

Noll, R.G., Older-Aguilar, D., Ross, R.R., & Rosston, G.L. (2001). The digital divide: Definitions, measurements, and policy issues. *Bridging the digital divide* (pp. 1-12). Sacramento, CA: California Council on Science and Technology.

OECD. (2000). *OECD small and medium enterprise outlook.* Paris.

Slowinski, J. (2000). Workforce literacy in an information age: Policy recommendations for developing an equitable high-tech skills workforce. *First Monday,* [Online], 5(7). Available: www.firstmonday.org

U.S. Census. (2001). *City and county data book.* Washington, D.C.: U.S. Census.

U.S. Census. (2003a). *U.S. census of population, 2000.* Washington, D.C.: U.S. Census. Available: www.census.gov

U.S. Census. (2003b). *County business patterns. Detailed NAICS classified data.* Washington, D.C.: U.S. Census. Available: www.census.gov

Warf, B. (2001). Segueways into cyberspace: Multiple geographies of the digital divide. *Environment and Planning B: Planning and Design, 28,* 3-19.

## Chapter IV

# The Ethics of Web Design:
## Ensuring Access for Everyone

Jack S. Cook
Rochester Institute of Technology, USA

Laura Cook
State University of New York, USA

## Abstract

*Web accessibility is really not a technological issue but rather a cultural problem. A Web site is said to be Web accessible if anyone, regardless of capabilities or disabilities, using any kind of Web browsing technology to visit the site has full and complete access to the site's content and has the ability to interact with the site if required. If properly planned from the start, a Web site can be functional, accessible and aesthetically pleasing. This chapter focuses on ensuring access to information available on the Internet. The overall objective is to increase awareness of Web accessibility issues by providing rationale for why Web designers should be interested in creating accessible sites. Specifically, this chapter identifies some of the emerging digital barriers to accessibility encountered by those with disabilities. Current efforts to address these barriers legally are identified and their effectiveness for breaking down barriers is discussed. The World Wide Web Consortium's (W3C's) Web Accessibility Initiative (WAI) is discussed, followed by a study of the 50 most visited Web sites. The chapter concludes with a discussion of the importance of this topic and future developments.*

# Introduction

*"The power of the Web is in its universality. Access by everyone regardless of disability is an essential aspect."*

—Tim Berners-Lee, W3C Director and inventor of the World Wide Web

The Web has altered how many people work, learn, and even play. With publishing, electronic commerce, distance learning and delivery of governmental services, the Web has quickly become an integral part of our society. Some would even argue that Web access is mandatory for success in our information-driven society. Technology offers people of all abilities ways of interacting with the world that were previously unimaginable. The Web remains a ray of hope for full and equal access to information, goods and services. Unfortunately, it also erects barriers for those with disabilities.

Today, half of the world's population, 3 billion worldwide, are either disabled or in direct contact with the disabled. According to some accounts, America is home to 50 to 55 million totally or partially impaired or disabled citizens who possess a wealth of ideas and are gifted. This estimate is stated as a range since it is ever-changing. Anyone can become disabled through injury, illness, or aging. In fact, most of us will experience at least some period of temporary or permanent disability as a result of life's unpredictability. A broken arm, stroke, moderate hearing loss, or repetitive stress syndrome can render someone disabled. Furthermore, circumstances can create a situation that mimics a disability, forcing someone to rely on the same solutions used by those with disabilities. If your mouse stops functioning, you must rely on your keyboard – the same way someone who is blind or quadriplegic might. Assessing a Web page in a noisy environment that makes hearing difficult increases the importance of audio transcripts just as these transcripts are important to someone with a hearing impairment.

There are three major accessibility issues encountered by people with disabilities: (1) access to computers in general, (2) access to Web browsers that support the needs of those disabled, and (3) access to the documents available on the Internet and WWW. This chapter focuses on the third category. The overall objective is to increase awareness of Web accessibility issues by providing rationale for why Web designers should be interested. Specifically, this chapter identifies some of the emerging digital barriers to accessibility encountered by those with disabilities. Current efforts to address these barriers legally are identified and their effectiveness for breaking down barriers is discussed. The World Wide Web Consortium's Web Accessibility Initiative (W3C's WAI) is discussed, followed by guidelines. These guidelines do not discourage content providers from using multimedia but rather explain how to make it and other content more accessible to a wider audience. In addition, a study of the 50 most visited Web sites is explored. The chapter concludes with a discussion of the importance of future research and applications requiring further study.

# The Facts

It is estimated that the disabled community in the U.S. has $1 trillion in disposable income (Tillett, 2001). There are approximately 2.5 million people in the U.S. alone that consider themselves blind of which 600,000 to 800,000 are legally blind according to the National Federation of the Blind (Wingfield, 1999). Melanie Brunson, Director of Advocacy and Governmental Affairs at the American Council of the Blind estimates that only 10 to 20% of all Web sites are accessible to the blind (Brown, 2000). Only 10% of those disabled regularly go online, whereas 30% of the rest of the population do so (Alexander, 2000). A recent Harris poll found that those with disabilities that access the Internet spend twice as much time online and using e-mail as compared to people without disabilities (Solomon, 2000). Sadly, professional opportunities for these individuals have been limited. Seventy-one percent of the 17 million people nationwide with disabilities of working age are jobless, according to the Able to Work Consortium, a non-profit advocacy group for the disabled (Alexander, 2000). According to the U.S. Department of Labor, the percent is even higher for working-age blind people — 74% (Zielinski, 2000). Individuals with disabilities who are employed earn on average one-third the income of those who are not.

In this age of skilled labor shortages, it is wise to examine this relatively untapped work force, particularly when the only barrier to hiring a competent, excellent employee may be a piece of inexpensive hardware or software. Also, those with disabilities often are eligible for third party funding based on medical necessity. Medicaid, insurance companies, vocational rehabilitation centers, and others usually will cover the cost of hardware and software for those with disabilities since it is less expensive to purchase the technology they need to function unassisted than it is to provide assisted care and disability benefits.

# The Problem:
## Awareness, Understanding, Knowledge & Sensitivity

Clearly, disabled Americans constitute a very large minority in this nation and the vast majority of the Web has arbitrarily excluded them. Disabled surfers find as much frustration online as they do information. The multimedia aspects of the World Wide Web which entices, excites and makes the Web useful to so many, erects barriers for others. Most Web sites would never intentionally exclude such a large group of people. Table 1 includes a sample of disabilities that impact Internet surfing behavior.

*Table 1. Sample of disabilities that impact surfing behavior*

| Visual Impairment | Hearing Impairment |
|---|---|
| Total Blindness<br>Tunnel Vision<br>Color Blindness<br>Age Related Visual Degeneration | Deaf<br>Hard of Hearing |
| **Cognitive/Learning Disabilities** | **Motor Disabilities** |
| Dyslexia<br>Dysgraphia (inability to produce handwriting reliably)<br>Attention Deficit Disorder | Quadriplegics<br>Arthritis |

Many Web content providers are unaware of the problem (lack of awareness). The unaware often wrongly believe that the disabled are only a small percentage of the population (lack of understanding). When confronted with the idea of making their Web pages accessible for the blind for example, some providers react as if you asked them to create a hairbrush for the bald (lack of knowledge). Many designers of Web sites and software applications are young people whose eyesight is 20-20, with dexterous fingers who believe that everyone loves and uses a mouse since they do (lack of sensitivity). Hence, much of the problem is educating Web designers that the problem exists, that it is in their best interest to make sites accessible to everyone and training them how to create accessible sites.

Accessible design does not mean that pages need to be text only. In fact, just about everything one sees on the Web could be made accessible while maintaining all the dynamic and visually rich content. Web accessibility is really not a technological issue but rather a cultural problem. A Web site is said to be Web accessible if anyone, regardless of capabilities or disabilities, using any kind of Web browsing technology to visit the site has full and complete access to the site's content and has the ability to interact with the site if required. Even if Web developers callously disregard the ethical imperative of Web accessibility, it is in their best interests to create accessible Web pages.

# Literature Review

The literature surrounding Web accessibility can be broken down into the following generic categories: (1) what is Web accessibility, (2) why is Web accessibility worthwhile, (3) how to make Web sites accessible, (4) impact of Web accessibility on higher education and libraries, and (5) how accessible are various types of Web sites. The number of articles in these different areas is ever-increasing and the importance of this topic is gaining momentum, especially within higher education and library science.

Many Web sites address the questions of what is Web accessibility, why is it important and how can a Web site be made accessible. Informative sites that address these three questions include Web Accessibility in Mind (http://www.webaim.org/), World Wide Web Consortium Web Accessibility Initiative (http://www.w3.org/WAI/), U.S. Government's Section 508 site (http://www.section508.gov/), HTML Writers Guild Web Accessibility Standards (http://hwg.org/opcenter/policy/access.html), and IBM Web Accessibility Center (http://www-3.ibm.com/able/) to name just a few. There are also sites that publish statistics about accessibility. UsableNet publishes their Web Usability Index, which "is a freely accessible Web usability statistics database. It enables web site owners to compare their sites with industry averages and helps Web designers avoid the most common usability problems" (UsableNet, 2003). While usability is not the same as accessibility, this site looks at both issues. Donston (2003) looked at the issue of why this topic is important in the article "Web Access for All: Accessibility Issues Too Important to Ignore". With respect to costs, it was found that the "total cost of making a site adhere to 508/WAI guidelines is estimated at between $180,000 and $200,000, including testing and continuous monitoring, according to Meta Group Inc. research, while the cost of making accessibility part of a Web site design process will be about $45,000 to $50,000" (Donston, 2003, p. 54).

Higher education seems to be at the forefront with respect to accessibility. Some universities, such as the Rochester Institute of Technology's College of Information Technology, teach Web accessibility in their curriculum. In addition, librarians and students in library science are strong proponents of Web accessibility. Cheryl Riley states, "The library profession has long championed making materials readily available to all patrons. Indeed, providing materials and information to everyone equally is one of the fundamental beliefs inherent in our profession" (Riley, 2002). While increasingly more individuals are aware of it, are they really doing anything about it at the college level? Schmetzke's (2001) research looked at the 24 highest ranked schools from *US News & World Report* of library and information science. The department Web site and main library site were evaluated for each school. Schmetzke found that only 23% of the department Web pages were Bobby approved and 59% of the library main pages were Bobby approved (Schmetske, 2001). Another study in this general category was conducted by Lilly and Van Fleet (1999). They evaluated (using Bobby) the main library page of the "100 most wired colleges" of 1998 according to Yahoo! They found 40% of the pages were accessible (Lilly & Van Fleet, 1999).

In the category of how accessible Web sites are, many articles are written with specific disabilities in mind. For example, articles in the journal *Library High Tech* written by Riley (2002) and Axtell and Dixon (2002) are both aimed at looking at Web accessibility and technologies associated with the visually impaired. In terms of data gathered about how accessible certain sites are, many studies used small sample sizes (100 sites or less). More research is needed to address the larger picture of how inaccessible most sites are.

# Ensuring Access Helps Everyone

Many people enjoy accommodations and devices originally designed to address problems encountered by individuals with disabilities. Many of these are easily recognizable such as ramps, elevators, Braille, closed-captioning and much more. The purpose of these tools is to make common facilities more accessible to the disabled population. In addition, these features have also been used and desired by everyone, especially families with strollers who can take advantage of elevators and curb cuts. Devices such as speaker phones and vibrating pagers were created to address the needs of the disabled community. The absence of these simple yet effective features results in a direct violation of governmental regulations as well as human rights. Would you deem it morally acceptable to restrict the hearing impaired from watching television because television networks and manufacturers refused to implement closed-captioning? Most people would say no. How about the World Wide Web? Would you accept disabled persons not being able to utilize the resources of the Internet simply because they are disabled? Table 2 shows samples of situations that arise with non-disabled individuals that mimic Internet access problems encountered by those that are disabled.

There is good reason to believe designers that make their Web sites accessible will see similar benefits to their non-disabled customers online. Sighted surfers

*Table 2. Situations that mimic a disability*

| Visual Impairment | • Driving a car – eyes needed for another task <br> • Under- or over-illuminated rooms |
|---|---|
| Hearing Impairment | • Noisy environment |
| Cognitive/Learning Disabilities | • English as a second language (ESL) – automatic language translators exist but are not 100% reliable <br> • Illiterate |
| Motor Disabilities | • Small mobile devices <br> • No keyboard or mouse <br> • Hands-free environment |

with slow modems may choose not to display images if alternative text is provided. Furthermore, an increasingly growing number of mobile devices are Internet-enabled. Excitement exists in the disabilities community over wireless since Web pages designed to be more easily accessed by these wireless devices better serve the needs of the disabled — furthering both the goals of m-commerce and Web accessibility.

Accessible Web pages work well with PDAs and cell phones. Although many individuals that use mobile devices are not disabled, typing on tiny keypads and reading text on small screens (particularly while driving) creates an opportunity to combine complex voice-recognition technology with wireless Web-access services. One day, computers in cars will regularly download content from Web sites designed to be accessible by the disabled and output that content using computer-generated speech — saving the driver the hassle of pulling over and reading the content on an output screen (Tillett, 2001). So, in this case, a Web page designed with screen readers in mind would be easily accessible by both mobile devices that are voice-enabled and low speed devices.

# Digital Barriers to Accessibility

Web accessibility makes Web sites accessible by handicapped or disabled users. Most people would expect that accessibility would be designed to help the blind, visually impaired, or deaf. However, Web accessibility encompasses a larger group of disabilities. It also includes cognitive disabilities, physical disabilities, age-related visual degeneration, and economical disadvantages. The following sections detail how each of these disabilities are affected by poor Web design.

## Blind, Visually Impaired, Age-Related Visual Degeneration

Originally, the Internet was not laden with graphics and thus was highly accessible to the blind. The widespread use of "graphical user interfaces (GUIs), which relies heavily on visual cues for navigation, wasn't exactly cause for celebration in the blind community" (Zielinski, 2000, p. 42). The Web has become a very visual medium. If you cannot see, you would not have access to the content contained in images and videos. The group of visually impaired individuals is diverse in the varying degrees to which they can see. Some are totally blind. Other individuals suffer from tunnel vision, which is a visual impairment that only allows the eye to see a portion or narrow scope of normal view. There are individuals who suffer from color blindness. In addition, there are also those who suffer from age-related visual degeneration, which happens when older people have difficulty reading fine print.

One device designed for blind and visually impaired users to gain access to the Internet is a screen reader. A screen reader is the commonly-used name for Voice Output Technology that produces synthesized voice or Braille output for text displayed on the computer screen, which includes icons, menus, text, punctuation and control buttons as well as keystrokes entered on the keyboard. In addition, many screen readers cannot interpret special formatting, such as a change in color signifying a difference or importance or other visual cues that help a sighted person access information. Another problem faced by screen readers is that it reads the content of the Web page from left to right and then from top to bottom (Valenza, 2000). This is a problem when a screen reader is trying to read charts or columns in a tabular format. Figure 1 illustrates the problem when a screen reader reads text from a table on a Web page. The screen reader does not know that it should read each column or cell, which is something a sighted user would know. This again puts the blind or visually impaired user at a disadvantage.

Screen readers cannot interpret images on Web pages. Therefore, Web designers need to include text descriptions of pictures or links. This is very easy to do by using an ALT tag. Charts and pictures contained on Web pages explain information, but if you could not see the picture, you would lose that information. A screen reader cannot describe the picture unless there is a descriptive paragraph to explain the chart's information or what the picture is displaying.

*Figure 1. How a screen reader interprets a table*

| What You See on a Web Page: | | |
|---|---|---|
| | Mary had a little lamb. | Jack and Jill went up the hill. |
| The Screen Reader Says the Following: | | |
| | Mary had a Jack and Jill went little lamb up the hill | |

Many Web designers do include the ALT tag to label the graphics. However, they tend to just list the name of the picture or chart, which clearly does not give details about the graphic.

Also, a screen reader cannot interpret colors, which are often used to differentiate items. If a user is reading a tutorial on corporate finance, and red text signifies a wrong answer and green text signifies a right answer, the screen reader would only read what is written and not indicate that the answer was right or wrong.

Another adaptive technology for the visually impaired is screen enlargers or magnifiers. These devices enlarge the text for individuals with low vision so that they can read the screen. The user determines the amount of the screen to be enlarged, so the user can enlarge an entire screen, a certain portion of the screen, or just the area surrounding the mouse (Ray, 1998). This technology is great because the visually impaired user can now interact with the online content rather than printing out the information in a larger font or Braille.

Since this technology is fairly new, it still has problems relaying some information to users. One disadvantage experienced is that when certain texts are enlarged, the words and sentences become fragmented and distorted. The user has to "mentally reassemble" the pieces to get a better understanding of the content he or she is reading (Ray, 1998). For some users, it is frustrating to remember what you just read and then connect it to what you are reading. The enlargers also distort the page so that the layout of the page is unclear and therefore creates difficulty when navigating the site. Another problem encountered is that some graphics cannot be magnified. With this problem, the user still cannot view the image, making that information inaccessible to them.

The two devices discussed, screen readers and screen enlargers, are only a small portion of the technology being created to make the Internet more accessible to the blind and visually impaired. They offer some hope to users in accessing information that was once closed to them. However, the barriers have not totally been broken down because Web designs must be changed to take advantage of these technologies.

## Deaf and Hard of Hearing

The deaf and hard of hearing population does not have as many problems interacting with the Internet. Their primary means of communication is through visual mediums. The Internet falls into this category. The major problem these individuals encounter when on the Internet is with the use of audio files and video files that include audio. If they are at a Web site that utilizes audio files, they cannot hear it and therefore, the information on the audio clips are inaccessible. The solution is "captioning". With captioning of audio files, deaf and hard of hearing individuals can see the words being displayed on the screen while the audio file is running. This solution enhances their interaction with Web sites, especially those that are based on music and advertising.

## Physical Challenges

Imagine being a quadriplegic who has battled cerebral palsy since birth. Attached to your head is a metal wand designed to press keys on a standard keyboard. As you surf the Net, you awkwardly press hot keys and keyboard shortcuts. You would like to use Microsoft's Internet Explorer or Netscape Navigator, but although keystroke commands exist in both browsers, their key positioning layouts are too difficult and tiresome — literally it is a neck-breaking task to surf. Web accessibility not only affects the auditory and visually impaired, but the physically and cognitively/psychologically handicapped find it is harder for them to surf the Web as well. The cognitively and physically disabled population consists of people that "may have difficulty reading long pages or understanding text" or those that "may not be able to move easily or use a keyboard or mouse when creating or interacting with the image" (Nielsen, 2000, p. 309). The targeted groups are the elderly, quadriplegics, dyslexics, and physically immobile users.

In order to accommodate the physically handicapped population, there are several accessibility programs or applications designed to help them view the Internet more effectively. For example, voice applications are used to enable surfers to browse the Internet using voice commands. In March 2000, the World Wide Web Consortium (W3C) developed a voice program called VoiceXML.

Although the physically impaired have the ability to complete various operations through a dialogue session using VoiceXML, there is still a need for software implementation. One of the disadvantages that the user faces is sequential formatting of the software. This can become an inconvenience for some when retrieving information. For example, the user could be looking up information on a search engine and receive a tremendous amount of search results. VoiceXML will not selectively give the user the highlighted areas.

Instead, the person will succumb to hearing the whole list before responding to the preferred request. So there is an urgent necessity for more advanced applications that will best fit the needs of impaired users. It is also important to design "sticky" keys that the disabled can hit in sequence when executing a command rather than simultaneously (Tillett, 2001). The CTRL + C for copy is an example. Those with restricted mobility should be allowed to hit CTRL followed by C rather than being required to hit both keys simultaneously. Emerging adaptive technologies do include foot-powered or head-controlled mice and keyboards.

## Cognitive Disabilities

The term "cognitive disabilities" does not refer to people with below-average intelligence. Rather, it refers to such disabilities as learning disabilities, Attention Deficit Disorder, brain injury, and genetic diseases such as Down's syndrome and autism (WebAim, 2003). Cognitive disabilities have not been the focus of as much human/computer interface research as physical disabilities. Providing those with cognitive or learning disabilities with multi-sensory experiences, interaction, and positive reinforcement enhances their experiences. While Web developers cannot necessarily make their pages completely usable by those with cognitive disabilities, there are some strategies in the design of the Web pages that will make it easier. For example, keep the navigation consistent and clear, simplify the overall layout, group and organize content into smaller sections, and supplement text with either illustrations or media (WebAIM, 2003).

## Economically and Technologically Disadvantaged

Economically disabled users are often overlooked for the reason that they are not seen as disabled like a blind person. When a blind person goes to use the Internet, they need special equipment as discussed previously. This Web accessibility issue deals with computer users who may use older technology, and therefore issues may arise such as modem speed.

More software designers are creating software designed to run on relatively new computer systems. If a user has an old computer that will not run scripts and applets including JavaScript and Shockwave, then they are to be considered disabled in that sense. In order to make the Internet accessible to every user, a designer needs to incorporate the idea that users can be economically disabled.

## Input and Output Needs

Table 3 summarizes some input and output needs for various disabilities.

*Table 3. Input and output needs*

| Disability | Input Needs | Output Needs |
|---|---|---|
| Learning | • Quiet work area<br>• Ear protectors<br>• Word prediction programs (software that predicts whole words from fragments)<br>• Spell checkers<br>• Thesaurus<br>• Grammar checker | • Large print displays<br>• Voice output |
| Hearing and/or Speech Impairments | • Standard keyboard<br>• Standard mouse | |
| Low Vision | • Standard keyboards with large print key-top labels | • Voice output system<br>• Special software that can reverse the screen from black to white or white on black for people who are light sensitive |
| Blind | • Standard keyboard<br>• Braille key labels for keyboards<br>• Braille input devices<br>• Scanners with optical character recognition | • Voice output<br>• Earphones<br>• Refreshable Braille displays<br>• Braille translation software and Braille printers |
| Illiterate | | • Voice output |

# Current Legal & Regulatory Efforts to Eliminate Barriers

In the highly competitive and largely unregulated software, hardware, and telecommunications industries, government action is needed to ensure those with disabilities are not forgotten. Free market forces alone are unlikely to address the needs of people of all abilities. Web accessibility is one area where rational regulation and legislation by the government is needed.

## Americans with Disabilities Act (ADA) of 1990

Public Law 101-336 enacted the Americans with Disabilities Act (ADA) on July 26, 1990. The ADA was the largest civil rights legislation since the Civil Rights Act of 1964. It covers any new construction of public accommodations and commercial facilities after January 26, 1993 under Title III, Reference 28 CFR 36. The ADA prohibits discrimination against persons with disabilities and provides a private course of action to enforce its provisions. It mandates that "public accommodations" make their goods and services accessible to people with disabilities, unless doing so creates "undue burden". Among other things, it mandates handicapped accessible parking with ample room between spaces and curb cuts, handicapped entrances with wheelchair ramps and oversized doors, handicapped toilets, as well as appropriate signage throughout a building.

Are Web sites legally obligated to accommodate the disabled, just as "public accommodations" such as shopping malls are under the ADA? Groups like the National Federation of the Blind (NFB), along with state and federal officials, maintain that the ADA requires Web sites to take reasonable steps toward Web accessibility. However, the law has traditionally been applied to physical places like restaurants, hotels, and retail establishments, not virtual places like Web portals. But the Internet is quickly transforming itself into the digital equivalent of a "public accommodation". For example, the blind benefit greatly from online grocery stores such as Homegrocer.com, a site acquired by Webvan (Tillett, 2001). An ongoing legal debate exists about whether the ADA applies to the Internet, which has at least generated discourse over whether Web sites need to be accessible. Lawsuits have been filed, but for now cases are being settled out of court. For example, the NFB, a 50,000-member advocacy group, filed suit against America Online (AOL) in November 1999. It alleged that AOL was not complying with the ADA since its service did not work with special screen reader software used by the blind. In July of 2000, the organization dropped its suit, citing progress AOL made on making its service accessible to the blind. AOL agreed to make future versions of its software compatible with screen readers.

The best indication that the ADA may apply to the Web came in a policy ruling dated September 9, 1996 (10 NDLR 240) from the Civil Rights Division of the U.S. Justice Department. It stated, "covered entities under the ADA are required to provide effective communication, regardless of whether they generally communicate through print media, audio media, or computerized media such as the Internet". For now, most companies are forced to act on Web accessibility. Roger Petersen, whose vision is impaired, worked with the California Council of the Blind to negotiate a settlement with Bank of America to get talking ATMs and a Web accessible online banking site (Solomon, 2000).

Medical transcriptionist Rose Combs is blind and surfs regularly. When she went online to vote in the Arizona Democratic primary, the first election held on the Web, her screen reader could not translate the Election.com Web site (Solomon, 2000). With assistance from the NFB, Combs successfully lobbied to make the Web site's online voting accessible to screen readers. Even though most companies are cooperative after being approached by the disabled community, Web accessibility needs a more proactive approach to ensure that the moat of inaccessibility does not prevent people like Combs from initially participating in such important activities as voting. Corporations considering offering Internet voting as an option for shareholder participation at annual or special business meetings should ensure accessibility issues are addressed.

## State Policies

Many states have recently enacted policies similar to New York State's Technology Policy 99-3. The policy requires all New York State agencies provide universally accessible Web sites to enable persons with disabilities to access the sites. While this policy was passed on September 30, 1999, there are still many state agencies that have not complied.

The Sutton Trilogy, three Supreme Court cases from 1999, narrowed the definition of a disability such that if there are mitigating factors, then a condition is not a disability from a legal perspective. For example, impaired vision may be corrected using eyewear. Where does this leave those with disabilities? John Kemp, a former employment lawyer and appointee to the National Council on Disability, as well as the senior VP for strategy development at HalfThePlanet.com (a Web portal designed for those with disabilities) is not sure (Alexander, 2000). He wears four artificial limbs and wonders legally if based on the Sutton Trilogy whether his prostheses are "mitigating factors" under the ADA. His artificial limbs allow him to walk and type much as eyewear allows those with impaired vision to see well enough not to be legally blind. Although obviously to the average citizen, John is disabled, according to the Supreme Court rulings in 1999, it is not clear to him whether he is legally disabled.

While some laws have emerged to make Web sites accessible, the Web is still growing at an unprecedented rate, with many Web developers totally unaware of the need to make sites accessible. In addition, many companies, primarily out of ignorance, are not interested in making their Web sites accessible. A few hours of surfing with your browser set to not display images as well as having Java and JavaScript disabled will quickly convince you of how far designers have to go before the Web is accessible. Governments are obligated to ensure all their constituents have equal access to information and services, the Web being no exception. Luckily, the federal government is attempting to change this through its enormous buying power.

## Section 508 of the Rehabilitation Act of 1973

Federal agencies spend more than $30 billion on IT products and services yearly, making them collectively the largest purchaser of IT in the world (Tillett, 2001). The executive branch of the government is not covered by the ADA but rather by the 1973 Rehabilitation Act as well as recent amendments. The Workforce Investment Act of 1998, which amended Section 508 of the Rehabilitation Act, requires federal agencies to make information technology accessible to those disabled (Tillett, 2001). It requires agencies to consider disabilities when buying or building technology (Web design tools, photocopiers, e-mail software, kiosks, keyboards, databases, servers). As these agencies demand technology from vendors that satisfy requirements of Section 508, companies such as IBM, Microsoft, and Oracle are incorporating accessibility features into their core products. Even though the private sector is not required to comply with Section 508, accessible technology will benefit everyone since it "would be ludicrous and overly expensive to create one product for the federal government and another, less accessible product for the private sector" (Tillett, 2001). The Access Board, a small, independent federal agency responsible for drafting and implementing the new accessibility standards issued recommendations that federal sites must meet by mid-2001. The

Congressional Budget Office estimates that the resulting improvements in productivity of the 168,000 disabled federal workers from implementation of Section 508 will generate savings of up to $466 million (Pressman, 2000).

Section 508 applies to all Web sites of federal agencies whether designed to be accessed from the Net or a government intranet. However, it does not apply to the private sector nor to recipients of Federal funds. The World Wide Web Consortium (W3C) promotes accessibility in both the private and public sectors.

# Web Accessibility Initiative (WAI)

The mission of the WAI: *The W3C's commitment to lead the Web to its full potential includes promoting a high degree of usability for people with disabilities. The Web Accessibility Initiative (WAI), in coordination with organizations around the world, is pursuing accessibility of the Web through five primary areas of work: technology, guidelines, tools, education & outreach, and research & development.* In addition to making their own Web site accessible, one way that private companies can support the W3C's WAI is to decline to advertise on sites that are not themselves Web accessible. IBM has taken this approach (Bray, 1999). As reported, it was not IBM's objective to cancel advertising but rather to get Web sites to comply with WAI.

To get an idea of how accessible the Web is, research was conducted on how accessible the 50 most visited Web sites are (Cook, Martinez, Messina & Piggott, 2003). In this study, the researchers replicated a study conducted by Terry Sullivan and Rebecca Matson. Their study was reported in "Barriers to Use: Usability and Content Accessibility on the Web's Most Popular Sites," published in 2000. Sullivan and Matson determined the 50 most popular sites through the use of www.Alexa.com. Accessibility of these sites may have either worsened or improved between the year 2000 and 2002, when these data were collected. The goal of this project was to evaluate whether or not and to what degree this had occurred. This research was significant because, as seen in this chapter, Web accessibility is a growing issue in human rights and should be addressed as technology progresses. The findings of the study are discussed next.

# How Accessible are the 50 Most Visited Web Sites?

The analysis portion of this project was completed by using the tool *Bobby* from http://www.cast.org/bobby. *Bobby* tests Web sites according to two criteria: (1) The W3C (World Wide Web Consortium) guidelines. This refers to fourteen guidelines that are outlined by the W3C in an attempt to evaluate Web content accessibility. See *http://www.w3.org/TR/WCAG10/* for the 14 guidelines. (2) The U.S. Section 508 guidelines. This refers to rules that pertain to Section 508 of the Rehabilitation Act of the ADA, which are closely knit with the Web content guidelines proposed by the WAI (Web Accessibility Initiative).

The *HTML validation tool* was also utilized to see if the Web pages conformed to HTML standards. This is significant because "another way to ensure that a web site has optimal accessibility is to use HTML to convey meaning and not format or layout" (Casey 1999, p. 23). Prior to evaluating the 50 most visited Web sites, it is necessary to discuss the priority levels created by the W3C. There are three levels, and each describes accessibility to a different degree.

- **Priority 1:** These checkpoints *must* be fulfilled. A failure to satisfy the Priority 1 guidelines implies that the page entails severe dilemmas with respect to accessibility for disabled individuals.

- **Priority 2:** These checkpoints *should* be fulfilled. A failure to pass this would deem the page as fairly accessible but would also highlight key access matters that should be addressed.

- **Priority 3:** These checkpoints *may* be fulfilled. "If you can pass all items in this section, including relevant User Checks, your page meets Conformance Level AAA for the Web Content Guidelines" (http://www.cast.org/bobby).

## Sullivan and Matson's Research

Sullivan and Matson evaluated the 50 most frequently visited sites of the year 2000.

## *Methodology*

The scope of the project entailed the evaluation of eight guidelines that were titled under Priority 1, corresponding to guidelines 1.1, 1.2, 2.1, 6.1, 6.3, 7.1, 9.1, and 11.4. The researchers analyzed only the main page of the sites. This was because it would have been exhaustive to check each page contained in the sites. Similarly, if the main pages were deemed inaccessible or poorly accessible, navigation through and onto other pages can be assumed to be similar. The researchers also evaluated the sites using automated tools and manually. The significance of this was to decipher between potential and actual failures. Sites that contained text-only alternative home pages were evaluated solely on that text version. In order to display the results of the project more adequately, a four-tier model was used. These were ranked as follows:

- High Accessibility. This refers to sites that contained no perceived content access problems.
- Medium Accessibility (Mostly Accessible). This refers to sites with few accessibility problems; five and less checkpoints identified.
- Medium Accessibility (Partly Accessible). This refers to sites with many accessibility problems; between 5-10 checkpoints identified.
- Inaccessible. This refers to sites that contain major obstacles in the use of the page; more than 10 checkpoints identified.

## *Sampling*

Sullivan and Matson (2000) took a sample of 50 Web sites. These sites were significant because they represented elements of a purposive sample. That is to say, the sites were categorized as frequently visited. Therefore, they should be reasonably accessible to everyone.

## *Findings*

Table 4 (Sullivan & Matson, 2000, p. 142) shows the results of their findings. The results were formatted in a four-tier table according to the rankings previously defined. Sullivan and Matson constructed the table in order to rank sites from highly accessible to inaccessible. The sites in the first tier are in

*Table 4. Accessibility of the 50 most visited Web sites in 2000*

| Tier 1<br>Highly Accessible | Amazon, Gohip, Google, Goto, Hotbot, Microsoft, Monster, MSN, Snap |
|---|---|
| Tier 2<br>Mostly Accessible | AltaVista, Att.net, Excite, Icq, Tripod, Geocities, Lycos, Angelfire, Iwon, Yahoo, Infospace, Go |
| Tier 3<br>Partly Accessible | Dogpile, Looksmart, Preferences, Xoom, Bluemountain, Ebay, ZDNet, Netscape |
| Tier 4<br>Inaccessible | 100free, Mp3, Homestead, Quicken, Ancestry, Webshots, Real, MSNBC, Freeserve, Cnet, About, Cnn, AOL, Hitbox, Askjeeves, Networksolutions, Ragingbull, Ign, Weather, Cdnow, This |

alphabetical order because their accessibility ranking was identical. For tiers two through four, the sites are listed in order of which was more accessible.

# Replication of Sullivan and Matson's Research

## *Methodology*

This research (Cook, Martinez, Messina & Piggott, 2003) entailed an evaluation of the 16 checkpoints that are titled under Priority 1; as stated previously, these guidelines must be satisfied in order to consider a site accessible. The purpose of focusing mainly on Priority 1 guidelines is not only to more easily draw similarities to the research conducted by Sullivan and Matson but also to maintain consistency. Priority 1 guidelines also represent the minimum requirements to be *Bobby* approved. In an attempt to be as accurate as possible, the researchers ranked the sites' accessibility according to actual issues encountered by *Bobby,* which excludes user-checks. "User checks are triggered by something on the page; however, you need to determine whether they apply" *(www.cast.org/bobby)*. These are usually checked manually because they represent potential, not actual violations. The significance of this is that the scope of the research solely deals with the automated evaluation of the Web pages excluding user checks. This research furthered the previous research of Sullivan and Matson (2000) by not only incorporating Web Content Accessibility Guidelines 1.0 but also the U.S. Section 508 guidelines under *Bobby* and the *HTML Validator* in order to be equipped to construct more detailed comparisons. As with the previous study, only the home page of each site was analyzed.

Each Web site was ranked in a similar four-tier model. However, the research technique of assigning the Web sites with a specific rating varied due to the manner in which these sites were tested. The sites were ranked according to:

- Highly Accessible. This includes sites that reveal no instances of Priority 1 guideline violations.

- Mostly Accessible. This includes sites that reveal only one instance of a Priority 1 guideline violation.

- Partly Accessible. This includes sites that reveal two instances of Priority 1 guideline violations.

- Inaccessible. This includes sites that reveal three or more instances of Priority 1 guideline violations.

## Sampling

Since this research was meant to be compared to the one formally conducted by Sullivan and Matson, this research attempted to utilize the same sites. Therefore the 50 most frequently visited sites of 2000, as obtained by *www.Alexa.com*, were again sampled. Due to the dynamic nature of the Web, a few of the Web sites were no longer valid for analysis in 2002 — three sites were no longer in existence and seven sites could not be successfully evaluated through *Bobby*. A possible reason is compatibility issues with JavaScript. In these cases, the researchers substituted those invalid sites with other sites that were relevant to the project scope. The substitute sites were frequently visited

*Table 5. Accessibility of the 50 most visited Web sites in 2002*

| Tier 1 Highly Accessible | FBI.Gov, Google, Microsoft, MSN |
|---|---|
| Tier 2 Mostly Accessible | AltaVista, AskJeeves, Att.Net, CDNow, CheapFares, Dogpile, Excite, Geocities, GoTo, HomeStead, Hotmail, ICQ, InfoSpace, Iwon, LookSmart, MTV, NetworkSolutions, Priceline, Quicken, Weather, WebShots, ZDNet |
| Tier 3 Partly Accessible | 100Free, About, Ameritrade, Ancestry, AngelFire, BlueMountain, BusinessNow, CNET, Ebay, Etrade, FreeServe, Go, GoHip, HitBox, HotBot, HotJobs, Tripod, Yahoo |
| Tier 4 Inaccessible | Amazon, EOnline, IGN, Lycos, MSNBC, Netscape |

sites in their respective categories during the year 2002, according to www.100hotsites.com. It is important to note that www.Alexa.com did not carry a list pertaining to the top most frequently visited sites at the time of the research; hence the reason this research used www.100hotsites.com to find substitute sites. The list of sites broken down by category is shown in Table 5.

## Findings & Interpretation

Generally, the researchers encountered negative results with respect to the overall accessibility of the Web's most frequently visited sites. The rankings of the 50 Web sites are represented in Table 5. Similar to Table 4, the sites are ranked from highly accessible to inaccessible. The listing is in alphabetical order for each category since the research methodology was based on a specific number of errors that needed to occur in order to be categorized into a tier.

Overall no sites were considered to be *Bobby* approved. However, 8% of the sites could be deemed as highly accessible since the analysis did not reveal any instances of Priority 1 violations. This implies that amongst these 50 frequently visited sites, 92% possessed an actual significant barrier with respect to accessibility. Eighty percent of the sites had medium accessibility, with 44% classified as mostly accessible, and 36% as partly accessible. The remaining 12% of the sites were considered inaccessible.

The sites were also evaluated according to the U.S. 508 guidelines. Among these 50 sites it was found that 96% of the sites were not 508 approved, while 100% of the sites did not pass the *HTML Validation Checker*. The U.S. 508 guidelines are less restrictive than those of the W3C. Hence, 4% of the sites that were not approved by *Bobby* were 508 approved.

Frequently violated guidelines were Priority Guidelines 1.1 and 12.1. The guidelines are described as follows:

- **Guideline 1.1:** "Provide a text equivalent for every non-text element" (Chisholm, Vanderhelden & Jacobs, 2001, pp. 39-40). Ninety-two percent of the sites violated this guideline at least once. This guideline can be fixed both easily and with a negligible amount of time.

- **Guideline 12.1:** "Title each frame to facilitate frame identification and navigation" (Chisholm, Vanderheiden & Jacobs, 2001, p. 46). Twenty percent of the sites violated this guideline at least once. In order to correct this problem, developers would have to name each frame appropriately, using the <title> tag in the HTML code to facilitate better accessibility.

The most frequently violated guidelines mentioned previously represent Web site design errors that can easily be altered to facilitate better Web accessibility for disabled individuals. Therefore, it would be appropriate to assume enforcement of regulations according to the W3C guidelines would significantly alter the accessibility of the Web today.

## Comparison of Research Studies

Table 6 compares the results of 2002 (Cook, Martinez, Messina & Piggott, 2003) with the previous study of Sullivan and Matson (2000). It depicts the specific guidelines that were violated, as well as their change, if any, over the two years. Included are also the sites that were replaced and their respective results.

Table 6 reveals many different trends. Nine sites were considered highly accessible during the year 2000; of these sites two were no longer valid for evaluation. Three sites, however, remained highly accessible in 2002. On the other hand, four sites moved into a lower tier classification. Therefore, of the seven working sites, 57% of these became less accessible between 2000 and 2002. The researchers suspect that the magnitude of the decline in accessibility

*Table 6. Comparison of accessibility studies*

| Highly Accessible (Year 2000) | Number of Priority 1 Errors (Year 2002) | Improved, Worsened, or Remained the Same? |
|---|---|---|
| Amazon | 3 (Guideline 1.1: 3 Instances) | Worsened |
| GoHip | 2 (Guideline 1.1: 1 Instance Guideline 12.1: 1 Instance) | Worsened |
| Google | 0 | Remained the Same |
| GoTo | 1 (Guideline 1.1: 1 Instance) | Worsened |
| Hotbot | 2 (Guideline 1.1: 1 Instance Guideline 12.1: 1 Instance) | Worsened |
| Microsoft | 0 | Remained the Same |
| Monster | N/A | N/A |
| MSN | 0 | Remained the Same |
| Snap | N/A | N/A |

*\*N/A means that the site no longer exists, or that there were errors when* Bobby *ran it.*

*Table 6. Comparison of accessibility studies (continued)*

| Mostly Accessible (Year 2000) | Number of Priority 1 Errors (Year 2002) | Improved, Worsened, or Remained the Same? |
|---|---|---|
| Altavista | 1 (Guideline 1.1: 1 Instance) | Remained the Same |
| Att.net | 1 (Guideline 1.1: 1 Instance) | Remained the Same |
| Excite | 1 (Guideline 1.1: 1 Instance) | Remained the Same |
| ICQ | 1 (Guideline 1.1: 1 Instance) | Remained the Same |
| Tripod | 2 (Guideline 1.1: 1 Instance Guideline 12.1: 1 Instance) | Worsened |
| Geocities | 1 (Guideline 1.1: 1 Instance) | Remained the Same |
| Lycos | 3 (Guideline 1.1: 2 Instances Guideline 12.1: 1 Instance) | Worsened |
| Angelfire | 2 (Guideline 1.1: 1 Instance Guideline 12.1: 1 Instance) | Worsened |
| IWON | 1 (Guideline 1.1: 1 Instance) | Remained the Same |
| Yahoo | 2 (Guideline 1.1: 2 Instances) | Worsened |
| Infospace | 1 (Guideline 1.1: 1 Instance) | Remained the Same |
| Go | 2 (Guideline 1.1: 2 Instance) | Worsened |

| Partly Accessible (Year 2000) | Number of Priority 1 Errors (Year 2002) | Improved, Worsened, or Remained the Same? |
|---|---|---|
| Dogpile | 1 (Guideline 1.1: 1 Instance) | Improved |
| Looksmart | 1 (Guideline 1.1: 1 Instance) | Improved |
| Preferences | N/A | N/A |
| Xoom | N/A | N/A |
| Blue Mountain | 2 (Guideline 1.1: 1 Instance Guideline 12.1: 1 Instance) | Remained the Same |
| Ebay | 2 (Guideline 1.1: 1 Instance Guideline 12.1: 1 Instance) | Remained the Same |
| ZDNet | 1 (Guideline 1.1: 1 Instance) | Improved |
| Netscape | 3 (Guideline 1.1: 2 Instances Guideline 12.1: 1 Instance) | Worsened |

| Inaccessible (Year 2000) | Number of Priority 1 Errors (Year 2002) | Improved, Worsened, or Remained the Same? |
|---|---|---|
| 100Free | 2 (Guideline 1.1: 2 Instances) | Improved |
| Mp3 | N/A | N/A |
| Homestead | 1 (Guideline 1.1: 1 Instance) | Improved |
| Quicken | 1 (Guideline 1.1: 1 Instance) | Improved |
| Ancestry | 2 (Guideline 1.1: 2 Instances) | Improved |
| Webshots | 1 (Guideline 1.1: 1 Instance) | Improved |
| Real | N/A | N/A |
| MSNBC | 3 (Guideline 1.1: 2 Instances Guideline 12.1: 1 Instance) | Remained the Same |
| Freeserve | 2 (Guideline 1.1: 2 Instances) | Improved |
| CNET | 2 (Guideline 1.1: 2 Instances) | Improved |
| About | 2 (Guideline 1.1: 2 Instances) | Improved |
| CNN | N/A | N/A |
| AOL | N/A | N/A |
| Hitbox | 2 (Guideline 1.1: 1 Instance Guideline 12.1: 1 Instance) | Improved |
| Ask Jeeves | 1 (Guideline 1.1: 1 Instance) | Improved |
| Network Solutions | 1 (Guideline 1.1: 1 Instance) | Improved |
| Raging Bull | N/A | N/A |
| IGN | 3 (Guideline 1.1: 2 Instances Guideline 12.1: 1 Instance) | Remained the Same |
| Weather | 1 (Guideline 1.1: 1 Instance) | Improved |
| CDNow | 1 (Guideline 1.1: 1 Instance) | Improved |
| This | N/A | N/A |

*Table 6. Comparison of accessibility studies (continued)*

| 10 Sites Added for Consistency | Number of Priority 1 Errors (Year 2002) |
|---|---|
| FBI.Gov | 0 |
| Priceline | 1 (Guideline 1.1: 1 Instance) |
| MTV | 1 (Guideline 1.1: 1 Instance) |
| Hotmail | 1 (Guideline 1.1: 1 Instance) |
| CheapFares | 1 (Guideline 1.1: 1 Instance) |
| HotJobs | 2 (Guideline 1.1: 1 Instance<br>Guideline 12.1: 1 Instance) |
| ETrade | 2 (Guideline 1.1: 2 Instances) |
| BusinessNow | 2 (Guideline 1.1: 2 Instances) |
| Ameritrade | 2 (Guideline 1.1: 2 Instances) |
| EOnline | 3 (Guideline 1.1: 3 Instances) |

can be attributed to the move towards graphically rich pages. The accessibility trends of the highly accessible sites of 2000 are shown in Figure 2. This figure illustrates the specific number of sites (in their respective tier), whose ranking either changed or remained the same in comparison to the previous research conducted by Sullivan and Matson.

*Figure 2. Accessibility trends of the highly accessible sites from 2000 to 2002*

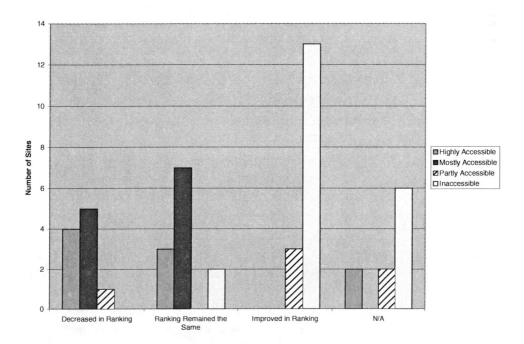

The sites that were classified as mostly accessible in 2000 had similar trends to those that were highly accessible in that same year. All 12 of these sites were valid for the 2002 evaluation. Of these 12 sites, seven remained in the mostly accessible bracket. Of the sites where accessibility was considered to have changed between 2000 and 2002, 100 percent fell into a lower tier of accessibility.

There were eight partly accessible sites of 2000; two of these were no longer valid for the 2002 appraisal. Two of these sites remained in the partly accessible classification while four shifted to other brackets. Three of these sites improved their accessibility. However, one site's accessibility level slipped into a lower classification.

The benchmark project conducted in 2000 ranked 21 sites as inaccessible. Of these sites, six were no longer valid for the 2002 evaluation. Two sites remained inaccessible to disabled users, while 13 sites moved into different brackets. These sites showed a general improvement concerning accessibility, revealing that 86% of the valid sites improved as opposed to remaining relatively constant. However, it is important to note that these sites still had a considerably higher level of accessibility problems in 2002.

## Limitations of the Study

One limitation of this study was that only the main page of each Web site was checked. Further research should check deeper into the site to see if other levels have the same level of accessibility. As mentioned previously, some Web sites no longer existed that were examined in 2000. In addition, the analysis tool *Bobby* has some limitations. According to the article "Evaluating Web Resources for Disability Access" (Rowan, 2000), some limitations of *Bobby* include: (a) the results may be overwhelming due to their length and complexity, (b) manual inspection may be required in order to separate actual problems from potential barriers, (c) may deem an acceptably accessible site as inaccessible, and (d) does not offer any recommendations for making a particular site more accessible. In addition, the user checks with *Bobby* leave some room for interpretation. There are other sites and tools that check accessibility. Cross-checking between multiple accessibility checkers would be helpful. They do not all analyze the same way, so data from multiple sources would create a broader picture of a site's accessibility or lack thereof. It is still essential that designers check their sites. Automated checkers cannot detect all accessibility problems.

Finally, with the ever-changing nature of the Web, any research data gathered at a single point in time are quickly outdated.

# Conclusion

It is easy for those of us who lead unencumbered lives to forget what it must be like to be denied daily activities that we take for granted — to be relegated to the sidelines of society. Software developers who create software that makes it easier for individuals to create Web pages should have requirements that elements are built into their software so that when code is created for Web pages, the code is Web accessible. Similar to requiring elevators be put in all new buildings, the Web needs to be accessible to everyone. Fortunately, with wireless devices gaining popularity, these devices will push technology to become more accessible to all because they require much of the same type of coding as required by assistive technologies. Luckily for those with disabilities, it is not a matter of if an organization will address the issue of accessibility, but rather when. Ultimately, Web accessibility needs to be a state of mind; not a set of guidelines.

# References

Alexander, M. (2000, September 4). Hidden workers: People with disabilities are missing from the new economy. *The Industry Standard, 160.*

Axtell, R., & Dixon, J. (2002). Voyager 2000: A review of accessibility for persons with visual disabilities. *Library Hi Tech, 20*(2), 141.

Bray, H. (1999, May 13). Making Web accessible. *Boston Globe, 1-2.*

Brown, D. (2000, June 5). Net to get more accessible to disabled. *Interactive Week, 32.*

Casey, C. (1999, March). Accessibility in the virtual library: Creating equal opportunity Web sites. *Information Technology and Libraries, 18*(1), 22-25.

Chisholm, W., Vanderheiden, M.G., & Jacobs, I. (2001, July/August). *Web content accessibility guidelines 1.0,* 35-53. University of Wisconsin, Madison.

Cook, J., Martinez, W., Messina, D., & Piggott, M. (2003, May) How accessible are the 50 most visited Web sites? *Proceedings of the 2003 Information Resources Management Association International Conference,* Philadelphia, PA, 75-78.

Donston, D. (2003, May 19). Web access for all; Accessibility issues too important to ignore. *eWeek, 20*(20), 54.

Joch, A. (2000, August 28). IE trips up disabled – but help is on the way. *eWeek,* 54, 60.

Lilly, E.B., & Van Fleet, C. (1999). Wired but not connected: Accessibility of academic library home pages. *The Reference Librarian,* 67/68, 5-28.

Nielsen, J. (2000). *Designing Web usability* (pp. 296-311). Indianapolis: New Riders.

Pressman, A. (2000, May 8). Carrot and stick. *The Industry Standard,* 120-121, 125.

Ray, D.S., & Ray, E.J. (1998, November). Adaptive technologies for the visually impaired: The role of technical communicators. *Society for Technical Communication,* 573-579.

Riley, C. (2002) Libraries, aggregator databases, screen readers and clients with disabilities. *Library Hi Tech, 20*(2), 179.

Rowan, M., Gregor, P., Sloan, D., & Booth, P. (2000). Evaluating Web resources for disability access. *ACM,* 80-84.

Schmetzke, A. (2001). Web accessibility at university libraries and library schools. *Library Hi Tech, 19*(1), 35-49.

Solomon, K. (2000, July 3). Disability divide. *The Industry Standard,* 136-137.

Sullivan, T., & Matson, R. (2000). Barriers to use: Usability and content accessibility on the Web's most popular sites. *ACM,* 139-144.

Tillett, L.S. (2001, February 12). Web accessibility ripples through IT: Tech mandate to aid disabled benefits broader audience. *Internet Week,* 1, 44-45.

UsableNet, Inc. (2003). What is the Web Usability Index? [Online]. Available: *<http://www.usablenet.com/wui/wui_index.html>*

Valenza, J.K. (2000, Fall). Surfing blind. *School Library Journal,* 34-36.

WebAIM. (2003). Types of cognitive disabilities [Online]. Available: <*http:/ /www.webaim.org/techniques/cognitive/*>

Wingfield, N. (1999, November 5). Must Web sites be accessible to the blind. *Wall Street Journal,* B1.

Zielinski, D. (2000, September). The age of access. *Presentations,* 40-51.

Chapter V

# Web Accessibility for Users with Disabilities:
## A  Multi-faceted Ethical  Analysis

Alfreda Dudley-Sponaugle
Towson University, USA

Jonathan Lazar
Towson University, USA

## Abstract

*When designing information systems, it is important to consider the needs of users with disabilities, including those with visual impairment, hearing impairment, or color-blindness. This is especially important for designing Websites. It takes time and money to create or redesign Websites for easy access for users with disabilities. This is also known as Web accessibility. This chapter will discuss the ethical implications involved with Web accessibility. An ethical analysis of Web accessibility will be performed using a partial combination of two structured analysis approaches. It is the position of the authors of this chapter that Website development should consider accessibility for users with sensory and motor disabilities as an ethical requirement.*

# Introduction

It is always important to consider the needs of users when building an informational system (Hoffer, George & Valacich, 2002). One portion of the user population is users with disabilities. Accessibility means to make a system or building usable by a person with a disability. When applied to informational systems, accessibility means that an information system is flexible enough to be utilized by someone using alternative input and/or output methods. These alternative input/output methods are known as assistive technologies (Alliance for Technology Access, 2000). Assistive technologies include screen readers (where the output on the screen is presented as speech synthesis output), voice recognition, and Braille output. When an information system provides flexibility for those using assistive technology, this is called an accessible information system. There are specific guidelines for making accessible information systems and accessible Websites. An important question to consider is "Why make a Website accessible?" While Web accessibility is an important goal, making a Website accessible costs money, and therefore it is important to identify clearly the stakeholders, the various concerns, and the benefits. The purpose of this chapter is to perform an ethical analysis of Web accessibility.

# Web Accessibility

For a user with an assistive technology to fully utilize a Website, this site must be very flexible to meet different user needs. Only then can the Website be called an accessible Website. To make a Website accessible, all forms of multimedia (such as graphic, audio, and video) must be available in a textual format (Paciello, 2000). Site navigation must work without graphics, applets, or other plug-ins, because otherwise a user cannot get past the home page (Lazar, 2001). For that matter, all portions of the Website must be available to someone who can browse using only text (Paciello, 2000). There are other challenges in making a Website that is accessible. For instance, tables that are used for page layout, instead of for presenting data, can be problematic if not clearly marked. Since many users with disabilities create and apply their own style sheets (also known as cascading style sheets or CSS), a Web page must still work properly when an associated style sheet is turned off. Important data must also be presented in a way that is useful to users with color-blindness. For instance, a color pie chart would not be a sufficient presentation of data. The actual numerical data, in a text form, should also be presented along with the pie chart.

To assist with making a Website fully accessible, there are two major sets of design guidelines. These guidelines help turn the high-level concept of Web accessibility into specific ways of implementing Websites that are accessible. The World Wide Web Consortium, as a part of the Web Accessibility Initiative, has a set of design guidelines called the Web Content Accessibility Guidelines (WCAG) to specifically guide those who strive to create an accessible Website. The guidelines are split up into three priority levels—Priority Levels 1, 2, and 3. Priority Level 1 includes guidelines that are a must for accessibility. Priority Level 2 includes guidelines that are important for accessibility. Priority Level 3 includes guidelines that are helpful for accessibility. The idea is that Priority Level 1 guidelines are most important, while Priority Level 3 guidelines are least important. The specific guidelines are available at http://www.w3.org/WAI. Another set of guidelines comes from the U.S. Federal Government. These guidelines for Web design, as a part of the "Section 508" initiative, specify rules that are similar to WCAG Priority Level 1, but are a bit more stringent. More information about those guidelines is available at http://www.section508.gov.

Because many people (including designers) are not very familiar with the guidelines and how to apply them, a number of software testing tools have been introduced to assist those who want to make their Websites accessible. These software tools essentially take the existing sets of guidelines (either the Web Accessibility Initiative or Section 508) and apply the guidelines to a specific Web page, looking for possible problems. Tools currently out there include LIFT, InFocus, PageScreamer, A-Prompt, and BOBBY. These tools, while helpful, tend to have limited effectiveness, since some of the guidelines still require human expertise to determine if the page is accessible (Lazar, Beere, Greenidge & Nagappa, 2003). For instance, alternative text is required for any graphic, but if the alternative text is "graphic here," this would be considered as accessible by a software testing tool, while realistically, it would not be helpful at all for the user. The testing software can also not determine whether Web pages are accessible when the style sheets are disabled. While these tools are not perfect, any help is needed in improving the current level of accessibility. Dot.com companies, non-profit organizations, and even state governments continue to have Websites that are inaccessible (Ceaparu & Shneiderman, 2002; Lazar et al., 2003; Sullivan & Matson, 2000). This is problematic, because it means that the wealth of information available on the Web is not available to users with disabilities. Some of the Websites would require only minor changes to make them accessible, but part of the challenge is convincing designers of the need for accessibility (Lazar et al., 2003). Framing Web accessibility in an ethical analysis will be helpful to inform designers of the need for accessibility.

# Ethics of Social Responsibility and Justice

There is a philosophical viewpoint that argues that the law is the foundation for ethical application (Edgar, 2003). The two major philosophers that support this position are Hobbes and John Rawls. Hobbes' position is that the law is this basis for society and that there can be no society without morality/ethics. John Rawls' position is the concern for fairness and that this fairness can only be applicable in creating contracts with individuals (Edgar, 2003). Are societies without laws or contractual agreements also unethical? Can societies apply ethics without legal interference? One major argument against this viewpoint is that societies can exist without laws to govern them.

In applying this philosophical view to the issue of Web accessibility to all users, it would seem that the focus on providing Web accessibility to all users would be a social priority for the Web designer. This is not the case. In a recent survey conducted by Lazar, Dudley-Sponaugle and Greenidge (2003), the majority of Webmasters indicated that they considered ethical principles when creating their Websites. In the same survey, the respondents indicated that the majority of their Websites are still inaccessible to users with disabilities (Lazar, Dudley-Sponaugle & Greenidge, 2003).

# Ethical Methodology

There are numerous approaches to analyze the ethicality of an issue or dilemma. When making ethical decisions involving computer technology, computer scientists and information systems professionals may use a philosophical analysis approach. Philosophical analysis is a continuous process, which involves several types of actions. The major activities involved in philosophical analysis are stating a position, justifying a reason for the position, and critically examining the position (Johnson, 2001). "There is no template approach to philosophical analysis in dealing with ethical issues or dilemmas involving computer technology" (Dudley-Sponaugle & Lidtke, 2002, p. 123). There are, however, several theories that are common in philosophical analysis of these issues. This chapter will analyze the ethics of Web accessibility by using the following philosophical theories: Consequentialism, Deontology, and Kant's Categorical Imperative.

This ethical analysis will combine two structured analysis approaches. One approach involves a four-step analysis process (Kallman & Grillo, 1996) and another approach involves an ethical process that engages controversial issues (Brown, 1999). The following is how the ethical analysis will be performed:

1.   Identifying the major ethical dilemma

2.   Different views

3.   Listing the stakeholders

4.   Applying the ethics of consequence (consequentialism)

5.   Analyzing rights and duties (deontology)

6.   Applying the ethics of principle (Kant's categorical imperative)

## Identifying the Major Ethical Dilemma

To do an ethical analysis, it is important to identify the major ethical dilemma, much as one would identify a research question. For this chapter, the major ethical dilemma is: *Should the needs of users with sensory and motor disabilities be taken into consideration in Website development?* We specifically identify users with sensory and motor disabilities because a lot more is known in the research literature about the technology needs of those user groups, as opposed to the needs of users with cognitive disabilities (e.g., mental retardation). It is the position of the authors that the needs of users with sensory and motor disabilities should become a standard consideration in Website development.

# Different Views on Web Accessibility

There are a number of different views involved when looking at the ethics of accessibility. There are three views that this chapter will address: economic, legal, and usability.

## *Economic*

If currently existing informational systems are re-designed for accessibility, there is a cost involved to make that information system accessible. In addition, for new information systems being developed, there may be additional expenses for testing the system for accessibility. For instance, testing a Website for accessibility could involve purchasing an assistive technology (such as *IBM Home Page Reader*), purchasing a software accessibility-testing tool (such as *PageScreamer* or *InFocus*), or hiring an expert to do an accessibility review. The costs of making an informational system accessible are offset by the possible economic benefits of an accessible system. If an organizational information system is accessible, it immediately increases the number of skilled workers who could use the system, making it easier to do hiring and find skilled workers. For a Website (such as an e-commerce site), the number of possible customers of the Website is immediately increased. An estimated 10-20% of people in the United States, Canada, Europe, and Australia have a disability (Paciello, 2000). Another estimate is that 13.1 million people in the United States use some form of assistive technology (Paciello, 2000).

## *Legal*

While there is currently no general requirement for accessibility of all informational systems, certain categories, in certain industries, are required to be accessible. For instance, the U.S. Government requires that all informational systems that the federal government develops or purchases be accessible to all individuals. This includes both organizational information systems as well as Websites. Some countries, such as Australia, are requiring certain categories of Websites to be accessible (Astbrink, 2001). Other countries are investigating the possibility of requiring certain categories of informational systems to be accessible. In the future, organizations with a minimum number of people (say, 100), or within certain industries may be required to be accessible (Lazar,

Kumin & Wolsey, 2001). A number of lawsuits have been filed in the United States against companies who have Websites that are not accessible. Probably, the best-known legal case was National Federation of the Blind vs. America Online (Lazar et al., 2001). As far as it is known, all of these cases have been settled out of court, which limits the power of these cases. Settling these cases out of court limits potential precedents that can be applied to other cases involving accessibility issues. However, the mere threat of a lawsuit may be considered by some organizations to be enough of an incentive to move their organization towards accessibility.

## *Usability*

Although this is not a traditional analysis, accessibility can be viewed through the lens of usability (ease of use). Many view accessibility as a subset of usability. That is, usability is general ease of use, while accessibility is ease of use for a specific population—users with disabilities. Many rules of general usability that are established in the literature (for instance, in Web design, limit animation, provide fast page downloads, etc.) are not frequently followed. However, a page that is more accessible (by having fast downloads, alternative text, text-based navigation, limited scripting) is also easier to use for the general user population (Lazar, 2001). A less complex user interface (which is better for users with disabilities) is also better for the general user population (Shneiderman, 1998). This has a similar parallel in the physical world—curb cuts help people with disabilities get from the sidewalk to the street level (Shneiderman, 2000). At the same time, many other people benefit from curb cuts: parents pushing strollers, people on bikes or roller skates, people pushing wheeled luggage, and others.

## Identifying the Major Stakeholders

Stakeholders are those individuals who have an investment in the question being decided. Each stakeholder may have different concerns and interests in the identified problem. Sometime these concerns and interests will conflict or compete. This means that the identified outcome of the decision may not be equally satisfactory to all stakeholders. The following individuals, groups, and organizations have been identified as being affected by Web accessibility:

- Individuals with disabilities (sensory and motor)
- Website developers
- Website businesses
- General public/users of the Website
- Government/legal communities

## Applying the Ethics of Consequence (Consequentialism)

Consequentialism (teleology) is the theory that an act (decision) should be determined by the effect that act (decision) has on others (Pence, 2000). There are three categories defined under this theory: egoism, utilitarianism, and altruism.

- Egoism: good for me, least harm to me
- Utilitarianism: good for the group, least harm for the group
- Altruism: good for all, some harm to me (Kallman & Grillo, 1996, p. 2)

In considering the way many currently existing Websites are developed with regards to the accessibility of users with disabilities, one could currently classify this under egoism. The best position to look at the ethical dilemma under consequentialism is from the utilitarian view. The utilitarian view looks at the best action for the group, which in this case would be access to the Web for all. In applying the consequentialism theory to our ethical dilemma, we want to answer the following questions:

*What will be the positive and negative consequences of the ethical dilemma?*

*Positive (Maximum Benefit for the Group)*
If the needs of the users with disabilities are taken into consideration, the action will benefit all the stakeholders. If the needs of the users with disabilities are not taken into consideration, the action will not benefit any of the stakeholders involved. The Website would benefit by increase in use from another segment of the population. There would be a decrease in the potential for costly lawsuits.

Consequentialists would agree with the decision to include users with disabilities in the design of Websites.

### Negative (Least Harm for the Group)

If the needs of the users with disabilities are taken into consideration, the action will result in the least harm for the Website developers. This harm is minimal because there are numerous software tools available to assist Website developers in making Websites accessible for users with disabilities. If the needs of users with disabilities were not taken into consideration, the result of this action would be harm not only to the individuals with disabilities but also to the other stakeholders.

The following is the analysis of consequentialism in a table format:

*Table 1. Analysis of consequentialism*

| Should the needs of users with sensory and motor disabilities be taken into consideration in website development? | Needs Taken Into Consideration (Yes) | Needs Not Taken Into Consideration (No) |
|---|---|---|
| Harmed | Website developers<br>Website businesses | Individuals with disabilities<br>Website developers<br>Website businesses<br>General public/users of the Website<br>Government/legal communities |
| Unharmed | Individuals with disabilities<br>Website developers<br>Website businesses<br>General public/users of the Website<br>Government/Legal communities | Website developers<br>Website businesses |
| Benefit | Individuals with disabilities<br>Website developers<br>Website businesses<br>General public/users of the Website<br>Government/legal communities | No Stakeholders<br>Identified |
| No Benefit | No Stakeholders<br>Identified | Individuals with disabilities<br>Website developers<br>Website businesses<br>General public/users of the Website<br>Government/legal communities |

## Which alternative will bring about the greatest overall good?

The alternative that will bring about the greatest overall good and least harm: The needs of users with disabilities should be incorporated in Web development.

# Applying the Ethics of Rights and Duties (Deontology)

"Deontological ethics holds that the rightness of an act is derived from its logical consistency and universalizability" (Pence, 2000, p. 14). Under this theory there are two main questions to be answered: What rights should be acknowledged, and what duties or responsibilities are not met?

## *What rights should be acknowledged?*

Users with disabilities have a right to access technology that is available to the general public. In the book entitled *The Right Thing to Do*, James Rachels list the following attribute, which he identified as important when viewing humanity from an ethical or moral point: " …the capacity to communicate, by whatever means, messages of an indefinite variety of types, that is, not just with an indefinite number of possible contents, but on indefinitely many possible topics" (Rachels, 1999, p. 102). To limit the Web to users without disabilities is to discriminate against someone simply because of his or her sensory or motor differences. These users have rights. The rights of users with disabilities have been acknowledged by many countries and international organizations (such as the UN). Their rights, as related to computer technology, have even been identified. However, this concept has not (as of yet) been fully expanded to include the Web environment.

## *What duties or responsibilities are not met?*

Deborah Johnson indicates in her book *Computer Ethics,* two concerns regarding Web accessibility: "On the upside, computer and information technology hold great promise for enabling the disabled…On the other hand, concern arises because the potential of the technology to help the disabled may not be pursued to the extent that it could be" (Johnson, 2001, p. 224). As part of the human community, it is the moral duty of the Website developer or creator to make his or her Websites available to persons with disabilities. Website developers who fail to consider users with disabilities accessing the Web contribute to a greater gulf in the digital divide. The digital divide is the lack of access to computer technology for certain individuals, due to economic and social disparity.

The following is the analysis of deontology in a table format:

*Table 2. Analysis of deontology*

| Should the needs of users with sensory and motor disabilities be taken into consideration in Website development? | Rights Abridged | Duties Neglected |
|---|---|---|
| Yes | X | X |
| No | • Individuals with disabilities (sensory and motor)<br>• General public/users of the Website | • Website developers<br>• Website businesses<br>• Government/legal communities |

In the deontology analysis, rights and duties correspond. This means that if rights are abridged then duties were neglected.

- **If the ethical dilemma were answered *yes;***
  - **The following stakeholders' rights might be abridged and duties neglected:**

    In this case, we identified no stakeholders. It should be noted that Website developers and businesses may experience their right to control their own property violated. But, in identifying corresponding neglected duties, it was awkward because the same stakeholders could be identified as neglecting duties.

- **If the ethical dilemma were answered *no;***
  - **The following stakeholders' rights might be abridged:**
    - *Individuals with disabilities (sensory and motor)*

      To deny these individuals the right to have access using this technology, based on a disability, would be violating their rights.

    - *General public/users of the Website*

      These individuals have a right to know the policies of the particular Website. If the public was aware, this could have an impact on future guiding principles and laws effecting Website development.

  - **The following stakeholders would be neglecting their duties:**
    - *Website developers and Website businesses*

      These stakeholders have a duty to create available access to all potential users of the Website. By ignoring the needs of users with disabilities, Website developers and businesses would be neglecting their duties.

- *Government/legal communities*

    These stakeholders would be neglecting their duties by not being diligent towards the protection of this segment of population in regards to access to Websites.

The deontologist position would indicate that it would be ethical to include the needs of users with disabilities into Website development.

## Applying the Ethics of Principles–Kant's Categorical Imperative

Immanuel Kant was an 18th century philosopher. [Kant] "…suggested two principles for examining whether a person has the right to act a certain way in a given situation…The principles of consistency and respect are aspects of what Kant called the Categorical Imperative" (Kallman & Grillo, 1996, p. 14).

### *The Principle of Respect: Are people treated as ends rather than means?*

To consider the users with disabilities in Web development indicates respect for those individuals. According to Kantian theory, not to consider the users with disabilities in Web development is an act of disrespect and violates the categorical imperative. To make a Website accessible shows respect to that user population (Kallman & Grillo, 1996).

### *The Principle of Consistency: What if everyone acted this way?*

This principle involves fairness in our actions to other individuals (Kaufmann & Grillo, 1996). It would not be good if everyone acted by ignoring the needs and wishes of certain users. The whole foundation of information systems design is to build systems for people. To build systems without consulting the users or even caring what is needed would make for unproductive systems and an unproductive society. Older users, younger users, and other diverse user populations have specific needs relating to computer technology, and these needs should be taken into consideration during the design of the technology.

# Conclusion

From our ethical analysis and previously mentioned studies, it is clear that making Websites accessible for users with disabilities is the ethical thing to do. Our analysis shows that by several different ethical theories it is unethical to exclude users with disabilities from Web accessibility. It is beneficial, legal, and economically sound for Websites to consider these individuals.

To help make the Web more accessible, there are two different areas that must be addressed:

## Political and Legal

There are two approaches for improving Web accessibility through political and legal means: statutory law, and case law. First, there must be a heightened awareness from governments around the world to protect the rights of these users from exclusion. Legal statutes that protect the rights of users with disabilities can go far to ensure equal access. Countries such as Australia, Canada, England, and the United States have all enacted accessibility laws that cover certain categories of Websites (Lazar, 2001). The second approach is to contact organizations and inform them of the inaccessibility of their Websites. Most organizations are simply unaware of the problem of inaccessibility. A last resort on this approach is to sue an organization. Lawsuits, such as the National Federation of the Blind vs. America Online, while settled out of court, force organizations to make their Websites more accessible (Lazar, Kumin & Wolsey, 2001). While there are no clear requirements in the United States for private companies to have accessible Websites, if a lawsuit of this nature did go to trial, it is possible that a precedent in the case law would be established, requiring accessibility for certain categories of private firms.

# Technical

While Web accessibility theoretically can be enforced by law, the question is how, from a technical point of view, to make a Website accessible. There are a number of different approaches for making a Website accessible. Obviously, it is easier to design a Website to be accessible in the first place, rather than "retro-fitting" a Website for accessibility at a later time. Web development tools, such as Macromedia DreamWeaver and Microsoft FrontPage, include tools to ensure that Websites, as they are being built, are accessible. In addition, automated accessibility software packages, such as InFocus, Pagescreamer, and A-Prompt, are available. These software tools can test already-built Websites, highlighting potential accessibility problems, and offering possible solutions.

The easier it is to make a Website accessible, the more likely that an organization will consider accessibility as an important goal. Unfortunately, the automated accessibility software tools can still be expensive and error-prone (Lazar, Beere, Greenidge & Nagappa, 2003). The guidelines for accessibility can be long and hard-to-understand for those without a background in accessibility. Both the tools and guidelines need to be refined and improved.

# References

Alliance for Technology Access. (2000). *Computer and Web resources for people with disabilities*. Berkeley, CA: Hunter House Publishers.

Astbrink, G. (2001). The legislative impact in Australia on universal access in telecommunications. *Proceedings of the 1st Conference on Universal Access in Human-Computer Interaction,* 1042-1046.

Brown, M. (1999). *The ethical process: An approach to controversial issues* (2nd ed.). Upper Saddle River, NJ: Prentice Hall.

Ceaparu, I., & Shneiderman, B. (2002). Improving Web-based civic information access: A case study of the 50 US States. *Proceedings of the IEEE International Symposium on Technology and Society,* 275-282.

Dudley-Sponaugle, A., & Lidtke, D. (2002). Preparing to teach ethics in a Computer Science Curriculum. *Proceedings of the IEEE International Symposium on Technology and Society,* 121-125.

Edgar, S.L. (2003). *Morality and machines: Perspectives on computer ethics* (2nd ed.). Sudbury, MA: Jones and Bartlett.

Hoffer, J., George, J., & Valacich, J. (2002). *Modern systems analysis and design* (3rd ed.). Reading, MA: Addison-Wesley.

Johnson, D. (2001). *Computer ethics* (3rd ed.). Upper Saddle River, NJ: Prentice Hall.

Kallman, J., & Grillo, J. (1996). *Ethical decision making and information technology: An introduction with cases* (2nd ed.). Boston: Irwin/McGraw-Hill.

Lazar, J. (2001). *User-centered Web development*. Sudbury, MA: Jones and Bartlett Publishers.

Lazar, J., Beere, P., Greenidge, K., & Nagappa, Y. (2003). Web accessibility in the mid-Atlantic United States: A study of 50 Web sites. *Universal Access in the Information Society, 2*(4), 331-341.

Lazar, J., Dudley-Sponaugle, A., & Greenidge, K. (In press). Improving Web accessibility: A Study of Webmaster perceptions. *Computers in human behavior*. Elsevier, Ltd.

Lazar, J., Kumin, L., & Wolsey, S. (2001). Universal usability for Web sites: Current trends in the U.S. law. *Proceedings of the Universal Access in Human-Computer Interaction 2001 Conference,* 1083-1087.

Paciello, M. (2000). *Web accessibility for people with disabilities*. Lawrence, KS: CMP Books.

Pence, G. (2000). *A dictionary of common philosophical terms*. New York: McGraw-Hill.

Rachels, J. (1999). *The right thing to do* (2nd ed.). Boston: McGraw-Hill.

Shneiderman, B. (1998). *Designing the user interface: Strategies for effective human-computer interaction* (3rd ed.). Reading, MA: Addison-Wesley.

Shneiderman, B. (2000). Universal usability: Pushing human-computer interaction research to empower every citizen. *Communications of the ACM, 43*(5), 84-91.

Sullivan, T., & Matson, R. (2000). Barriers to use: Usability and content accessibility on the Web's most popular sites. *Proceedings of the ACM Conference on Universal Usability,* 139-144.

## Chapter VI

# Internet Voting:
# Beyond  Technology

Trisha Woolley
Marist College, USA

Craig Fisher
Marist College, USA

## Abstract

*This chapter discusses issues relating to the use of Internet voting in public elections. Due to the recent problems in public elections and growing popularity of the Internet many are asking whether voting should take place via the Internet. Businesses are moving ahead in their use of the network to perform business transactions, implying that something as "simple" as counting votes could also be accomplished via the network. However, the United States is not ready to accurately and fairly count votes electronically from a sociological perspective. Analogies can be drawn between the distribution of the telephone and the distribution of the computer along various demographic lines that inform as to the viability of moving ahead too rapidly with I-voting. Local governments should continue to perform pilot voting projects that will pave the way for the future.*

# Introduction

One would think that by the 21st Century the most technologically advanced country in the world would be able to accurately count votes. However, the Presidential election of 2000 has led to outrage at the lack of ability to simply count votes. The state of Florida became the primary focus as its votes were counted and recounted. "An entire nation shared in a bug reporting exercise that will likely accelerate fundamental changes to how we administer democracy in the near future" (Weiss, 2001, p. 24).

Currently, voting takes place at supervised local polling sites with largely antiquated polling machines. Due to the recent popularity of e-commerce many are asking whether the voting process should take place electronically through the use of the Internet. Proponents of Internet voting (I-voting) suggest that I-voting will solve several financial and social problems (Cranor, 1996; Mohen & Glidden, 2001; Sink, 2000). An individual could vote from his/her home or office, rendering obstacles such as traffic, weather and working hours irrelevant. Disabled people and "shut-ins" could have easy access to voting systems (Sink, 2000). In addition, since computers can accurately and rapidly tabulate millions of financial transactions daily, the public naturally believes that I-voting may improve the accuracy of elections (Gugliotti, 2001), may increase voter turnout and is more secure than punch card systems (Raney, 1999).

In contrast to the optimists, there are several who highlight the technical difficulties (Phillips & Von Spankovsky, 2001; Rothke, 2001; Schwartz, 2000; White, 2001). In addition, I-voting departs from traditional voting techniques in that it uses computers that are "not necessarily owned and operated by election personal" (California Internet Voting Task Force (CIVTF), 2000, p. 3). This supervision is a cornerstone of our election process and to maintain principles of secret ballots and free elections, the United States government must approve all election equipment and procedures.

While I-voting is not yet approved for usage as election equipment, it is being tested and observed in various small-scale elections. Pennsylvania's Montgomery County moved from mechanical to I-voting, replacing its 40-year-old voting booths with new MicroVote machines in 1992. The March 2000 Arizona Democratic Party is the first time I-voting was used in a presidential preference primary (Mohen & Glidden, 2001). The U.S. Military conducted a small election (Phillips & Von Spakovsky, 2001) that illustrated that people can vote over the Internet under ideal conditions. However, it contained only 250 voters and most conditions are far from ideal.

This chapter reviews the issues and concludes that our nation is not yet ready for I-voting; however, the many advantages of I-voting, noted above, should not be lost. While there are many technical concerns with I-voting such as security, authentication, privacy, access, and data quality, there are many who believe that these can be resolved. The technical issues may be addressed and tested through an increase of local elections performed via the Internet gaining knowledge and experience (Mohen & Glidden, 2001). However, the technical issues may not be the biggest problem. The sociological problem of unequal distribution of equipment and capabilities along demographic lines causes a potentially serious degree of unfairness of I-voting (Kennard, 2002; Noll et al., 2002; Novak & Hoffman, 1998). This problem is discussed further in the section below entitled "Digital Divide".

# Background

In the U.S. Presidential Election of 2000 there was a major voting disaster in Florida that resulted in the denial to vote and under-counting issues due to poor equipment and operations. There were between 4 and 6 million votes not counted in Florida in the 2000 presidential election. Three million of the uncounted votes were lost to registration mix-ups, along with 500,000 to 1.2 million votes not counted due to voter error, archaic equipment, chaotic recount rules, and poor operations at polling places (Kennard, 2002).

The voters did not properly punch the holes in the bits of paper used, referred to as "chads". Every time the cards were fed through the reader the chads lined up differently or eventually broke off, resulting in different counts time after time. Voters who used out-dated punch card machines were seven times more likely to have their ballots discarded (Kennard, 2002). To compound matters, the ballots were also tearing as they were fed through the optical scanning machines, making them unreadable.

Ambiguous marks led to human error during the manual recount. The 4% error rate dramatically exceeded the allowable .001% margin in the Florida presidential election. "Thinking that human recounts with that sort of variation can check the accuracy of a machine count is rather like trying to recheck a machine's measurement of electron width using the human eye and a yardstick" (Elhauge, 2002, p. 18). Whenever there is ambiguity then there is room for inadvertent variation based on people's perceptions (Hall, 1991). Individuals performing the counting have their own political preferences that can result in a biased count.

Voters punched the wrong hole on the Butterfly Ballot due to a lack of uniformity and guidance. It has been estimated that 4,000 people voted for the wrong candidate believing the second hole represented the second candidate. In Duval County, "election officials disqualified 21,942 ballots because voters had chosen more than one presidential candidate" (Ruppe, 2000). Many voters were denied the right to vote because computers were down, not working, taking too long to warm up, or precincts closed their doors early. A lack of reliable databases in Florida allowed some ineligible voters to cast ballots, while keeping others who were eligible from voting.

The disaster that took place in Florida gave rise to the question of the ability to accurately count votes. The errors of the Florida 2000 election vote count gave realization that the current voting machines are not error-proof. To increase the accuracy when counting votes, the government must invest in new technology (Calmes, 2001; Gugliotta, 2001).

# Internet Voting

Modern information systems can process millions of financial transactions daily, perform e-commerce and implement a variety of remote devices for personal transactions such as ATM machines. So why not use information systems to perform the most fundamental task of counting? Ideally, obtaining official election results could be nearly instantaneous and perfectly accurate. There would be no human error during the counting of votes, increasing voting tabulation accuracy. Updating voter roles creates more accurate voter registration information, making it easier for election officials to keep track of voters.

There are several possible solutions to the Florida 2000 disaster and ways to improve our capacity to count votes. Replace infamous punch-card ballots with state of the art touch screens, train poll workers, or create a statewide registration database and modernize election laws. There are different types of voting systems, such as: punch card, mark-sense, direct recording electronic (DRE) machines, and I-voting. Due to the high popularity of the Internet we chose to examine Internet based voting systems and the use of computers to accurately count votes.

An Internet voting system is an election system that uses electronic ballots that can be transmitted to election officials over the Internet using computers not necessarily owned and operated by election personnel (California Internet Voting Task Force (CIVTF), 2000). I-voting is not considered electronic voting (or e-voting) that uses electronic kiosk machines in controlled locations such as polling stations. I-voting can be done at home, for example.

# Small Scale I-Voting Experience (Pilot Projects)

During 2000 several individual pilot projects performed I-voting in California, Pennsylvania, Switzerland, the Alaska Republican Presidential preference poll, the Arizona Democratic Presidential primary, and the National Reform Party primary. In addition, many college campuses are conducting student government elections via the Internet. We chose to focus on two, Arizona Democratic Primary elections in June 2000 (Mohen & Glidden, 2001) and the U.S. Department of Defense general presidential election in November 2000 (Faler, 2003), due to their relatively large size and scope.

The Democratic Primary elections in Arizona conducted a pilot that allowed voters the option to vote over the Internet. Media attention to this pilot was widespread. This internal party election applied party rules to voting. Internet based voting took place over four days while attendance based voting was held during the typical twelve-hour period (Mohen & Glidden, 2001).

The Federal Voting Assistance Program of the U.S. Department of Defense developed a trial of Internet voting for Defense personnel located outside the U.S. Four states made the necessary legislative changes to participate in the pilot (Dunbar, 2001). Only one county in each of the four states could participate, resulting in a total of 250 voters involved in the trial (Gibson, 2001/2002). The environment was strictly controlled, the infrastructure was impressive, and contractors were commissioned to develop the application and manage the technical environment (Barry et al., 2001).

Several pilot projects provide hope there is a technological answer to the counting problems faced by the 2000 election in Florida. However, the decision to implement Internet voting technology faces issues beyond counting, such as technological, societal, economical, and political issues. The election system voting requirements affected by these issues include: (1) it is not possible for a vote to be altered, (2) it is not possible for a validated vote to be dropped, and (3) it is not possible for an invalid vote to be counted in the final tally (Cranor, 1996).

The implication is that "better" voting machines are not machines better at counting, but better at correcting voter errors. Florida 2000 election failed due to bad interfaces, ambiguity in presentation and high tolerances for error in voters' input (Elhauge, 2002). The technological issues include security, authentication, access, and data quality. Societal issues address the digital divide, voter privacy, convenience, people with disabilities, public confidence, and the media. In addition, there are legal and economic issues.

# Security

A secure system is one in which only authorized voters can vote; no one can vote more than once; no one can determine for whom anyone else voted; no one can duplicate anyone else's vote; no one can change anyone else's vote without being discovered; and every voter can make sure that his or her vote has been taken into account in the final tabulation (Rothke, 2001).

Voting fraud is very real in any election, for example, "the 1997 Miami mayor's race was thrown out due to massive absentee ballot fraud" (Phillips & Von Spakovsky, 2001, p. 75). The term "dead people voting" refers to situations in which names of deceased individuals have not been taken off registration roles and absentee ballots are being requested by others who cast a vote in the deceased's name.

A secure system "is one that can withstand attack when its architecture (cryptography, firewalls, locks, etc.), is publicly known" (Rothke, 2001, p. 16). Typically, systems use simple user-ids and passwords, but these are considered risky because hackers can use software tools to discover most passwords (Rothke, 2001).

Some say that I-voting systems are more secure than punch card systems that require human intervention (Raney, 1999). The overlapping of several applications increases security, as in the case of the Arizona democratic election where system layers of user interface, business logic, and database access, combined with a third party count, adequately secured the votes (Mohen & Glidden,

2001). Independent review by knowledgeable experts and public observers is essential (Phillips & Von Spakovsky, 2001).

Viruses and denial of service attacks are easier to perform in the Internet environment than with traditional voting methods (Phillips & Von Spakovsky, 2001). Votes can be changed resulting from an attack on servers that are used to receive, collect, and store electronic votes. Threats to host computers have higher risk, are more detrimental in their outcome and are harder to detect than conventional threats. They are high risk because large numbers of votes could be manipulated at once without being detected (Mohen & Glidden, 2001). Security installed on home and office computers is usually minimal, which is "why viruses like the notorious 'Code Red,' which on July 19 infected more than 250,000 computer systems in just nine hours, are so successful" (Dunbar, 2001).

The distributed nature of I-voting makes it difficult to establish tampering patterns that are detectable (Weiss, 2001). The profile of these hackers is not just teenagers that are tech savvy, but powerful interest groups and foreign governments that want to change the outcome of the election to their best favor.

Any computer under an adversary's control can be made to simulate a valid connection to an election server, without actually connecting to anything. So, for example, a malicious librarian or cyber café operator could set up public computers appearing to accept votes but actually doing nothing with them (Rubin, 2002).

Threats to the Internet are compounded by the millions of novices who would be required to use the system (Phillips & Von Spakovsky, 2001) and poorly designed interfaces (Weiss, 2001). The help desk lines were overloaded during the Arizona pilot (Walsh, 2000). I-voting requires an infrastructure in which 200 million people could vote on a single day but no such system has yet been implemented. Some popular electronic funds transfer systems can only perform that many transactions per day, not per hour (Rothke, 2001). The combination of workload, inexperience and a new technology will result in a negative outcome.

# Authentication

User-ids and passwords are not just a security issue, but also an authentication issue that people really are who they say they are when they log onto the computer. Authentication of voters "ensures that every voter has the opportunity to cast a ballot and no voter is able to vote more than one time" (California Internet Noting Task Force, 2000, p. 1).

Vote selling is viewed as a major threat to I-voting. Large groups of voters may gladly sell their votes to the highest bidder if an Internet system is invoked and if there is no way to authenticate the voter. More sophisticated identification such as retina recognition is required by an I-voting system to verify both the identity and eligibility of potential voters, but identification software is not yet accessible to all voters (CIVTF, 2000). Today's machines do not offer options that would prevent accidental voting for the wrong candidate (Schwartz, 2000; Weiss, 2001).

Individuals may lose passwords, thus preventing the ability to vote. The Arizona Democratic Primary pilot contained instances in which people lost their PIN (personal identification number) or people attempted to vote using their partner's PIN (Rubin, 2001). The small scale of this pilot gives recognition to the difficulties in implementing this solution in a wider environment.

Authentication takes place when the voter logs onto the computer and is prompted with two questions and his or her assigned PIN. The PIN is similar to an ATM account number that banks and financial institutions use. It contains seven characters consisting of numbers and letters, resulting in two billion different combinations, making it very difficult to guess the PIN and match it to the individual to whom it was assigned. The two questions include information such as one's mother's maiden name, social security number, or birthday (Mohen & Glidden, 2001).

The system encrypts the vote and sends it through the electronic vehicle though which it undergoes double encryption, entailing security. The privacy of the voter's decision is upheld as the vote and ID are separated, never to meet again. The encrypted vote is then sent on a zip disk that only the auditor can read, at which point it is counted.

# Access

It is hoped that this technology inevitably increases access. Nevertheless, access to I-voting does not necessarily lead to increased voter turnout, and I-voting could actually be seen as a barrier. People are more likely not to vote due to apathy rather than because they cannot use their PCs (Ritchie, 2002). Out of the 250 eligible voters during the U.S. Department of Defense pilot project, only 84 actually voted over the Internet. The Alaska Republican Party used Internet voting in January 2000 to run a straw poll on presidential candidates and only 35 of the 4,430 Republican voters actually voted via the Internet (Dunbar, 2001). Only 4% of the 843,000 eligible voters voted by means of the Internet in the Arizona Democratic primary (Coleman, 2001). However, barriers to access could lead to a decrease in voter turnout. When voting from home, voters face a potential barrier to access if their computer breaks down or there is a loss of electricity. During the Arizona election, a "one hour outage occurred due to a hardware failure in a router" (Mohen & Glidden, 2001, p. 82).

It is common in the Internet industry to support certain browsers' models and versions, primarily for the sake of updating feature and functionality requirements and security. Several Macintosh users had problems casting votes online and were urged to vote at a physical polling place during the Arizona primary election.

Another barrier to access would include Denial of Service (DOS), in which a hacker floods the Internet during an election. In the Arizona Democratic primary the voting system deflected several DOSs. "Intrusion-detection software monitored activity on the voting network, detecting when unusual activity occurred and filtering it out, thus preventing it from interfering with the servers. We also configured the system's firewalls and external routers to minimize the effect of a distributed DOS attack" (Mohen & Glidden, 2001, p. 79). However, intrusion detection software slows service.

# Data Quality

A quality I-voting system depends upon the accuracy of the database of registration records (Phillips & Von Spakovsky, 2001) and a reliable secure workstation-network infrastructure. Computers offer the opportunity to provide an interface to the database that is tightly controlled, has less human intervention, more accuracy of voting tabulation, and improved timeliness (Gaboury, 2002). The Internet could maintain voter registration information to produce more accurate voter rolls (White, 2001).

# Voter Privacy

Democracy in the U.S. depends on the foundations of the secret ballot. In the Arizona election, the vote was separated from the voter ID to ensure the privacy of the voter's choice. Privacy is one of the most important aspects of voting for elections (Mohen & Glidden, 2001) but I-voting does not give voters enough privacy (Larsen, 1999). Family members and friends may be privy to personal identifiers needed to "secure" online voting and could either coerce or simply vote in place of individuals. In the 1994 election in Montgomery, Alabama, partisans "assisted" incapacitated and even comatose patients with their ballots (Binstock, 2001).

Network administrators using their networked office computers could change ballots. I-voting might encourage organized voter coercion by groups such as employers, churches, union bosses, nursing home administrator, and others (Phillips & Von Spakovsky, 2001). The "opportunity to approach a voter with a baseball bat, buying and selling votes, especially from the apathetic, greedy, or poor" is increased with I-voting (Weiss, 2001, p. 21).

# Digital Divide

The voting rights act of 1965 states that no voting qualification or prerequisite to voting, or standard, practice, or procedure shall be imposed or applied by any state or political subdivision to deny or abridge the right of any citizen of the U.S. to vote on account of race or color (Kennard, 2002; Noll, 2002; Voting Rights Act, 1965). Legally, every voter must have equal access to a computer and an equal right to vote, which is not the case. The civil rights movements centered on achieving equity of access to the ballot box. But once the vote was secured for minorities, devices such as literacy tests were often used to prevent minority voting. Given the disparity of Internet access, remote Internet voting represents a new-millennium version of a literacy test (Phillips & Von Spankovsky, 2001).

The *digital divide* "refers to differences in access to and uses of information technology that are correlated with income, race and ethnicity, gender, age, place of residence, and other measures of socioeconomic status" (Noll, Older-Aguilar, Ross & Rosston, 2002). Those who have Internet access and those who do not affect the equality of the Internet voting process. The digital divide is a challenge to I-voting because of the wide differences of availability of I-access based on demographics such as age, income, education, region, occupation and ethnicity. Only half of Americans have Internet access (USA Today, 2002).

On the "have" side of the divide are those with higher income (Kennard, 2002; Novak & Hoffman, 1998), with higher education (Novak & Hoffman, 1998), those who are white (Phillips & Von Spakovsky, 2001), those who are younger (CIVTF, 2000), and those living in the western region of the U.S. (CIVTF, 2000). Nationally, as of September 2001, only 40% of African Americans and 32% of Hispanics had Internet access from any location, compared to 60% of whites (U.S. Department of Commerce, 2002). African-American and Hispanic households are only 40% as likely as white households to have home Internet access (Phillips & Von Spakovsky, 2001). These statistics are referred to in Figure 1.

Individuals who are younger are considerably more capable with the computer, leaving a disadvantage to those who are older. Only 19% of the population of Americans over the age of 65 would support Internet voting even if it could be made secure from fraud (California Internet Voting Task Force, 2000). During the Arizona Democratic Primary election, 75% of voters between the ages of 18-35 voted by means of the Internet (Hixson, 2003).

*Figure 1. Digital divide*

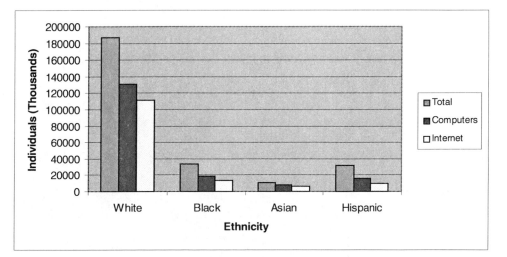

Those who are economically disadvantaged have less access to the Internet and thus would benefit little by I-voting. Household income explains home computer ownership. Increasing levels of income correspond to an increased likelihood of owning a home computer, regardless of race. Of families that have an average income of less than $25,000, only 28% had a computer and 19% had Internet access. Hispanic and Black households are reported as having median incomes of $33,565 and $29,470 respectively, while White households have median incomes of $46,305 and Asian and Pacific Islanders earn $53,635. Among individuals with incomes of $20,000 or less, whites are five times more likely to have Internet access than minorities (Phillips & Von Spakovsky, 2001).

Increasing levels of education correspond to an increased likelihood of work computer access, regardless of race. Whites are still more likely to own a home computer than African Americans at each and every education level, despite controlling for differences in education.

Voters in the western region of the U.S. have a higher degree of Internet proficiency and a higher degree of support for Internet voting at this time. During the Arizona Primary election, large urban counties with predominantly white voter populations vote via the Internet in much greater numbers than their counterparts in rural counties with large populations of minority voters. An analogy may demonstrate the significance of the *digital divide*. If public officials announced that they were going to add five more polling places within an all white neighborhood there would be outrage due to the inequity. But I-voting does just that – it puts a polling place in everyone's home who has I-access – just shown to be the "haves" (Phillips & Von Spakovsky, 2001).

Home Internet is the most rapidly accepted technology to ever have been deployed. There is a trend of new technologies being accepted faster with succeeding generations. For example, it took 38 years for the radio to be accepted and adopted by 50 million people and it took 30 years for the telephone to be accepted by as many people. Following the same trend, it took 13 years for the television and four years for the Internet to each be adopted by 50 million users (CIVTF, 2000). Over half of U.S. households have computers, of which 80% have Internet access. However, there are "6.3+ million American households that are still not connected to the most basic of telecommunications services: plain old telephone. Despite claims that basic telephone service is already universal, millions of Americans still are not connected to a network envisioned to reach from every home to every other home" (Educational Cyber Playground, 2000).

Computers today are more affordable to the economically disadvantaged than radios were in the past. Wal-Mart sells computers between $200-500 with Internet connection for less than $10 a month. Government subsidiaries and incentives would also lesson this divide. The increased existence of Internet cafes and kiosks lessons the digital divide among households and increases access. Children are using the Internet more and more. Eighty-one percent of girls and 74% of boys between the ages of 12-15 say they spend time online at least once a week (Gardyn, 2003). They will continue to do so as they age, thus decreasing the age digital divide.

## Convenience

A system is convenient if it allows voters to cast their votes quickly, on one session, and with minimal equipment or special skills (Cranor, 1996). The use of Internet voting may increase voter turnout for those living in remote locations, during bad weather conditions, or with hectic work schedules. The Arizona pilot increased actual voter turnout by more than 600%, from 12,844 in 1996 to over 86,000 in 2000, yet it could be attributed to the power of media attention (Election.com, 2000; Le Blanc et al., 2000). Those individuals that are between the ages of 18-25, who are busy students and professionals, and who do not find time to participate in elections, are also the most Internet savvy segments of the population (CIVTF, 2001). However, this could be an increased convenience for those that already have decided to vote. Over one 100 million people who were eligible to vote did not do so during the Presidential Election of 2000 (Rothke).

Obstacles are made irrelevant, as one could "vote in pajamas" and cast a vote from anywhere. This reduces travel time, the expense of gas, weather restrictions, time spent in lines, and minimizes distraction from normal activities, such as work. The user interface has the ability to be user-friendly. Double votes would not be allowed, eliminating the "chad" and butterfly ballot problem faced during the 2000 Presidential election. "You cannot overvote like we're seeing in Florida," said Bayless. "If you vote twice for one office, it will refuse to accept your ballot" (Copeland & Verton, 2000).

# People with Disabilities

I-voting can accommodate disabled people to cast votes (Mohen & Glidden, 2001). "McManus reports that in one large Florida county roughly 5% of all absentee ballots in a primary election, and 4% in a general election, were cast by residents of nursing homes and assisted living facilities" (Binstock, 2001, p. 132).

The computer interface can allow for several different fonts, languages, prints, and sizes to accommodate different voters. For example, larger font size can aid the visually impaired. Customizable interface and Internet access to the home will give disabled voters increased access to the voting process (CIVTF, 2000). However this potential is not always realized, as in the Arizona primary election where auditory prompts were omitted, greatly hindering blind voters (Phillips a& Von Spakovsky, 2001).

# Public Confidence

The Florida election has minimized trust in the current election process system, increasing the need for well-developed standards and platforms. The comfort level of voters (especially older individuals) affects the use of the Internet technology. The public must feel comfortable with the security, results, and outcomes in order to trust the technology (CIVTF, 2000). Sixty-three percent of adults currently oppose I-voting due to uncertainty of cyberspace (Phillips & Von Spakovsky, 2001). Many pilot tests involving thousands of people are needed prior to conducting an election with over 200 million voters.

In November 2000, voters in San Diego and Sacramento counties were able to try online voting from computers at polling places. The test was conducted for the state by VoteHere.net, which ran a similar trial in Maricopa County, Arizona. The company released polling results after the vote suggesting that "100% of voters who used the system found it easy to use, that 80% said they preferred I-voting to the current system, and that 65% said they would vote from home if they thought the system was secure" (Schwartz, 2000, C1).

However, if there are technological errors, there will be long lines, voter frustration, and loss of confidence in the election process. For example, during the Arizona 2000 presidential primary, "The Internet Corporation for Assigned

Names and Numbers (ICANN) board suffered from voter registration problems as well as overloaded servers that caused many voters to be turned away from the voting Web site" (Cranor, 2001).

# Media

Internet elections will "generate an enormous media interest" (Phillips & Von Spakovsky, 2001, p. 78). "Any real or perceived threat to a voter's privacy will probably lead to extensive negative publicity...When the technology driving teledemocracy fails, it is hard or impossible to cover it up. Such failures are widely publicized. The Monroe case was highly publicized for its failures...Even sub percentage failure rates may affect the votes of thousands of people. Such errors may, in addition to receiving much attention from the press, leave potential voters disillusioned about their role in our democracy" (Larsen, 1999, p. 57). Also, "it would be difficult to prevent political advertising from appearing on-screen and in the ballot window during voting if the voter's Internet Service Provider is one that displays advertising" (CIVTF, 2000, p. 26).

# Legal

State and federal laws are not geared toward governing remote electronic elections and the vendors that run them. There are no laws, standards or requirements for hardware or software. Legally, "Internet Voting opportunities must be accessible to all voters, including low income voters whose only access to the Internet may be through public access Internet terminals that are commonly available in libraries and schools. Internet ballots must be available in multiple languages in jurisdictions required to print multi-language ballots to conform to the Federal Voting Rights Act of 1965" (CIVTF, 2000, p. 5). Rothke (2001) said, "No electronic voting system is certified (even at the lowest level) of the US government". In addition to accessibility requirements, a lack of standards could lead to ballots that are as confusing as the butterfly ballot used in Florida (Rothke, 2001, p. 17).

# Economical

It is still not clear if I-voting is economically feasible. There are advantages to I-voting, but there are also disadvantages. The advantages include efficiency of administering elections and counting votes (Rothke, 2001), accuracy and speed of the automated voting system (Schwartz, 2000), availability of information and reduced transaction costs (Watson & Mundy, 2001), reduced number of polling places needed (Phillips & Von Spakovsky, 2001), and reduced travel expenses (Cranor, 1996).

An I-voting system would require several changes to the current system. This includes the cost of help desk support (Rothke, 2001), education of voters, training of election officials (Cranor, 1996), and reconfiguring of computer systems (CIVTF, 2000). An I-voting platform would require immense funding; for example Harris County in Houston, Texas is making the switch from punch card to electronic systems valued at thirty million dollars (Calmes, 2001). The U.S. Department of Defense pilot project cost the government $6.2 million, or $73,809 per vote (Dunbar, 2001).

Since voting is only performed once or twice per year, the market for voting software systems is relatively small (CIVTF, 2000). Therefore, election system vendors are forced by competitive bidding pressures to offer the cheapest possible systems with minimal fraud protection (Saltman, 1998).

# Recommendations

Corrective measures to ensure authentication include fingerprint and retina recognition. Many more pilots would need to be conducted with a technical support line to increase knowledge related to the issues surrounding Internet based voting. The government has expanded the pilot program initially performed by the U.S. Department of Defense to include ten states and 100,000 Americans overseas to cast ballots over the Internet in next year's election. (Faler, 2003). Accenture, a large consulting service, has said that it will develop an Internet voting system for the Defense Department to allow thousands of citizens to vote from abroad in the 2004 elections. The company will provide secure electronic balloting and voter registration for military personnel and their dependents and for some other United States citizens who live overseas who have a Windows-based computer and Internet access (*New York Times,* 2003).

The phased approach is recommended, starting on a small scale and then widening to include remote Internet ballots. Starting on a smaller scale provides for the use of Internet voting technology in a supervised setting like a traditional polling place. In this phase, voters would not yet gain the advantage of voting from any place at any time, but the integrity of the voting and tabulation technology will be verified through the use of Internet voting machines. This will help make the voting process more secure and standardized.

Moving to a larger scale will allow voters to cast remote Internet ballots. The authentication of voter identity would take place with a combination of manual and electronic procedures that would provide at least the same level of security as the existing voting process.

# Future Trends

The technology will not be an obstacle in the future of I-voting. Rapid advances in all areas of computing and networking are very promising. Mobile computing devices allow individuals to obtain Internet connection from virtually anywhere through the use of cell phones, laptops, and palmtops. These devices have become dramatically cheaper and much easier to use during the last few years. It is very common to see individuals talking on cell phones, using palm held computers and interacting with the Internet throughout the day — most college students do not remember "not having computers".

"Joseph N. Pelton, who directs the University of Colorado graduate telecommunications program compares the challenge of telephone voting with that faced by automatic teller machines a few decades ago. People said, 'I'll never use that.' Now people find them quite user friendly. It took quite a while to work out the bugs. That scenario may happen here" (Jacobsen, 1992, B1). Now, no one is surprised by the advances in technology. Specific examples included increases, in the three years 1994-1996, of PC ownership by 51.9%, modem ownership by 139% and e-mail access by 397% (McConnaughey et al., 1997). The trend is easily demonstrated that we are becoming an automated and information based society. It is a matter of time for the digital divide in computers and Internet to close as it has closed in radios and telephones.

And as we showed earlier, the population use or "take up" of new technologies is increasing at an exponential rate. While it took decades for radios and telephones to catch on, it has taken less than a decade for the Internet to mushroom in use. A recent government sponsored survey demonstrated that as a nation, "Americans have increasingly embraced the Information Age through electronic access in their homes" (McConnaughey et al., 1997).

The new wave of government will be a type of electronic-democracy (e-democracy) in which "town meetings" can be conducted on a national scale and numerous services of all types will be provided via the Internet. "Information technologies are being applied vigorously by governmental units at national, regional, and local levels around the world" (Marchionini, 2003, p. 25). The United States government has thousands of Websites, and many links to additional Websites. One recent report showed that 68 million Americans have used these Websites. Currently 98% of the schools in the U.S. and well over half the homes have Internet access (Marchionini, 2003). Americans can begin participating directly in making laws, says Jane Fountain (2003). Several examples are given in which citizens participated directly in establishing laws. With advances in the Internet the questions of application are only left to one's imagination. Questions such as should we go to war, decisions on federal spending, social programs, taxes, and numerous additional public issues can be addressed by the population as a whole. It is just a matter of time.

# Conclusion

While on the surface, I-voting appeared to be imminent; it now looks like it will be quite some time before we can depend on national elections over the Internet. The benefits, such as increased convenience, accuracy, efficiency, enhanced information and access to the disabled, are outweighed by the negative factors, such as inexperienced vendors, users, and election personnel, the digital divide, no laws for compatibility, privacy issues, and technological issues. Pilot tests should be conducted at local levels to facilitate integration, cooperation, and compatibility of the technologies, vendors, and users.

I-voting requires numerous technical and procedural innovations to ensure accurate voter authentication, ballot secrecy and security. Any socially respon-sible use of the Internet for voting purposes should be phased in gradually to ensure that election officials and members of the public are experienced, educated and confident with the technology.

The digital divide may always exist but in different forms. At some point all people may have Internet access just like they have the telephone but the quality of access might be different. For example, the digital divide might change to who has broadband versus who uses dial-up access.

# References

Barry, C., Dacey, P., Pickering, T., & Byrne, D. (2001, March). *Electronic voting and electronic counting of votes; A status report* [Online]. Available: http://www.eca.gov.au/reports/electronic_voting.pdf

Berghel, H. (2000). Digital village: Digital politics 2000. *Communications of the ACM, 43*(11), 17-23.

Binstock, R.H. (2001, February). Older voters in Florida and elsewhere: Butterflies, bingo, chads, and much more. The Gerontologist, *41*(1), 130-134.

California Internet Voting Task Force. (2000, January). (California Secretary of State). *A report on the feasibility of Internet voting* [Online]. Available: *www.ss.ca.gov/executive/ivote/*

Calmes, J. (2001, April 30). Broken ballot: America's dysfunctional voting system. Talk of voting machine overhauls is heating up a niche market. *Wall Street Journal,* A20.

Coleman, K. (2001). *Internet voting: Issues and legislation.* CRS Report for Congress [Online]. Available: http://164.109.48.86/topical/rights/democracy/inetvot110701.pdf

Copeland, L., & Verton, D. (2000, November 27). Arizona, California pilot voting over the Internet. *Computer World* [Online]. Available: http://www.computerworld.com/governmenttopics/government/*policy/story/0,10801,54409,00.html*

Cranor, L.F. (1996). Electronic voting: Computerized polls may save money, protect privacy. *Crossroads, 2*(4) [Online]. Available: *http://www.acm.org/crossroads/xrds2-4/voting.html*

Does the Internet Represent the Future of Voting? (2002, April). *USA Today, 130* (2683), 1-2.

Dunbar, J. (2001, August). Internet voting project cost Pentagon $73,809 per vote. *The Public i*[Online]. Available: http://www.notablesoftware.com/Press/JDunbar.html

Elhauge, E. (2001, December/2002, January). *The lessons of Florida 2000.* Policy Review, 15-37. Washington, D.C.

Faler, B. (2003, July). U.S. expands overseas online voting experiment. The Washington Post. Washington, D.C.

Fountain, J.E. (2003, January). Prospects for improving the regulatory process using e-rulemaking. *Communications of the ACM, 46*(1), 43-44.

Gaboury, J. (2002, January). The mouse that roars. *IIE Solutions. Norcross, 34*(1), 6.

Gardyn, R. (2003, April). Born to be wired. American Demographics, *25*(3), 14.

Gibson, R. (Winter 2001/2002). Elections online: Assessing Internet voting in light of the Arizona democratic primary. Political Science Quarterly, *116*(4), 561-584.

Gugliotta, G. (2001, July 17). Study finds millions of votes lost; Universities urge better technology, ballot procedures. *The Washington Post,* A01.

Hall, R.H. (1991) *Organizations: Structures, processes and outcomes.* Englewood Cliffs, NJ: Prentice Hall.

Hixson, C. L. (2003, December 2). *Voting for the new millennium: An implementation of e-voting* [Online]. Available: http://www.stetson.edu/departments/mathcs/students/research/cs/cs498/2003/chrisH/proposal.pdf

Jacobson, L. (1992, August 12). Telecommunications: Let your fingers do the voting, maybe. *Wall Street Journal* (Eastern edition), B1.

Kennard, W.E. (2002, March 7). Democracy's digital divide. *Christian Science Monitor,* 17.

Larsen, K.R.T. (1999, December). Voting technology implementation. *Communications of the ACM, 42*(12), 55-58.

Le Blanc, J., Wilhelm, A., & Smith, C. (2000, March 30). Arizona "ahead of its time" in online voting? *The Digital Beat, 2*(27) [Online]. Available: *http://www.benton.org/publibrary/digitalbeat/db033000.html*

Marchionini, G., Samet, H., & Brandt, L. (2003, January). Digital government. *Communications of the ACM, 46*(1), 25-27.

McConnaughey, J.W., Lader, W., Chin, R., & Everette, D. (1997).*The digital divide: A survey of information 'haves' and 'have nots' in 1997* [Online]. Available: http://www.ntia.doc.gov/*Ntiahome/net2/ falling.html*

Minority vote in Arizona Presidential Preference Primary strengthened by high voter turnout. (2000, March 24). *Election.com* [Online]. Availability: *http://www.election.com/uk/pressroom/pr2000/0324.htm*

Mohen, J., & Glidden, J. (2001, January). The case for Internet voting. *Communications of the ACM, 44*(1), 72.

Noll, Older-Aguilar, Ross & Rosston. (2003) *Bridging the digital divide: Definitions, measurement, and policy issues* [Online]. Available: http://www.ccst.us/cpa/bdd/BDDreport/BDD05.html

Novak, T.P., & Hoffman, D.L. (1998, February). *Bridging the digital divide: The impact of race on computer access and Internet use.* Vanderbilt University. Available: *http: //ecommerce.vanderbilt.edu/ research/papers/html/manuscripts/race/science*

Phillips, D.M., & Von Spankovsky, H.A. (2001, January). Gauging the risks of Internet elections. *Communications of the ACM, 44*(1), 73-86.

Raney, R. (1999, May 3). Casting ballots through the Internet. *New York Times,* C4.

Ritchie, K. (2002, January 8). Letter: A low poll for Internet voting. *The Guardian*, 17. Manchester, UK.

Rothke, B. (2001, Spring). Don't stop the handcount: A few problems with Internet voting. *Computer Security Journal, 17*(2), 13-21.

Rubin, A.D. (2002, December). Security considerations for remote electronic voting. *Communications of the ACM, 45*(12), 39.

Rubin, D. (2001, March). *The security of remote online voting* [Online]. *Available:* http://www.cs.virginia.edu/~evans/theses/rubin.pdf

Ruppe, D. (2000). Why Florida recounts favored Gore. *Abcnews.com.*

Saltman, R.G. (1998). Accuracy, integrity, and security in computerized vote-tallying. *Communications of the ACM, 31*(10).

Schwartz, J. (2000, November 27). E-voting: Its day has not come just yet. *New York Times,* C1.

Sink, M. (2000, November 2). Electronic voting machines let disabled choose in private. *New York Times,* G7.

Technology briefing software: Accenture to develop voting system. (2003, July 8). *New York Times,* C6.

U.S. Department of Commerce. (2002, February). A nation online: How Americans are expanding their use of the Internet [Online]. Available:http://www.ntia.doc/gov.ntiahome/dn/html/anationonline2.htm

Voting Rights Act of 1965. (1966). *South Carolina v. Katzenbach appendix.* Prentice Hall Documents Library. Available: *http://hcl.chass.ncsu.edu/garson/dye/docs/votrit65.htm*

Walsh, T. (2000, April). Arizona primary draws thousands of Internet voters in spite of officials' security concerns. *Government Computer News, 6*(4).

Watson, R.T., & Mundy, B. (2001, January). A strategic perspective of electronic democracy. *Communications of the ACM, 44*(1), 27-31.

Weiss, A. (2001). Click to vote. *netWorker, 5*(1).

White, B. (2001, March 7). Internet voting: A Web of intrigue?; Study says there's too much risk. *The Washington Post,* A21.

You have to have a telephone. (2000. *Educational Cyber Playground* [Online]. Available: *http://www.edu-cyberpg.com/Teachers/telephone.html*

## Chapter VII

# Protection of Minors from Harmful Internet Content

Geoffrey A. Sandy
Victoria University, Australia

## Abstract

*The Internet provides access to speech both conventional and unconventional. Some speech is considered harmful to minors. This chapter discusses the important social issue of how to best protect minors from such speech without violating the free speech rights of adults. It examines the Australian experience, one that has relevance to other relatively open societies like those found in North America and Europe. It concludes that the Australian regulatory framework has limited success in protecting minors from harmful Internet content and it risks compromising the free rights of adults.*

# Introduction

This chapter discusses the important social issue of how to best protect minors from harmful Internet content without violation of the free speech rights of adults (and minors). It examines the Australian experience, one that has relevance to other relatively open societies like those found in North America and Western Europe. Recently, interest was re-awakened on this important social and ethical issue with the publication by The Australian Institute study on youth exposure to sexual material on the Internet (Flood & Hamilton, 2003). It reported on a Newspoll telephone survey of 200 16-17 year olds, the ease of access, and the frequency of unwanted and wanted exposure to pornography based on gender.

In this chapter the Australian regulatory framework for the Internet is first described. Then the key questions concerning the protection of minors from harmful Internet content are discussed. Next the controversies and problems, together with their recommended solutions (protection strategies), are discussed. Finally, the main issues are restated and conclusions drawn with reference to the Australian experience. Note the term minors and children are used interchangeably throughout. Also note the abbreviations used throughout are COA (Commonwealth of Australia), CSIRO (Commonwealth Scientific and Industrial Organisation), EFA (Electronic Frontier Australia), OLFC (Office of Film and Literature Classification), HH (House Hansard), SH (Senate Hansard) and SSC (Senate Select Committee on Information Technology).

In order to address these matters a literature review was conducted and a detailed analysis of primary source documents was undertaken. The literature review concentrated on Australian sources but also included other countries, especially the United States. The most important primary source documents analysed were:

- Official Committee Hansards of the federal and state government hearings of submissions concerning the relevant bills.
- The House and Senate Hansard (and the state/territory equivalents) of the bills to regulate the Internet.
- The bills of the federal and state/territory parliaments to regulate the Internet.

- Official media releases, communiques and reports of regulatory bodies including the Attorney General's Departments.

In addition, documents published by the main non-government opponent of regulation—Electronic Frontier Australia (EFA)—and the government and regulatory bodies' response to the EFA are also analysed. These are particularly valuable in documenting the effectiveness of the regulatory framework in protecting minors from harmful Internet content.

# Australian Regulatory Framework

The Commonwealth of Australia is a federation of states (and territories). The Internet regulatory powers of the federal government are derived from the constitutional power over interstate telecommunications. The Internet regulatory powers of the states are more comprehensive and include authorship, access and distribution of Internet content. Australia has a long history of regulation of offline media and a concern to protect children from harmful material. It is an elaborate system to regulate the production, distribution and consumption of offline and online material for both adults and children. Some material, for instance, child pornography, is illegal for all. Some material is considered suitable only for adults.

Australia uses legislative classification systems to determine the type of regulation that is to be applied to the offline media like films, videotapes, publications and computer games. Internet content is treated like a film for classification purposes. However, the Internet regulatory body is the Australian Broadcasting Board (ABA). Television is "self regulated" but uses a similar classification to films. Films and videotapes, whether they are locally made or come from overseas must be classified before they can be sold, hired or shown publicly in Australia. A "Refused Classification" (RC) renders the film or videotape illegal. The Classification Board of the Federal Office of Film and Literature Classification (OLFC) undertakes classification of film and other media according to the relevant classification system. When making its classification decisions the Board is required to reflect contemporary standards and must apply criteria that are set out in the National Classification Code (the Code). The Code is determined under the Classification (Publications, Films and Computer Games) Act 1995. The Code expressly states that classification decisions must be based on a number of principles and particular attention should be given to protection of minors from material likely to harm or disturb them (OLFC, 1999). Each classification category contains a list of criteria used by the Board when making classification decisions. These criteria relate to the classifiable elements of violence, sex, coarse language, adult themes, drug use and nudity. In considering each element, the Board makes classification decisions based on a number of considerations including the impact of individual elements and their cumulative effect.

In making classification decisions the Board also decides what consumer advice should be provided. Australian law requires that consumer advice be shown with the classification symbol on posters, advertisements and video jackets. It is done to assist people to make informed choices about the films and other media they choose for themselves or for their children. One classification relevant to films and videos and also for Internet content is Restricted (R). The R category is legally restricted to adults and R material is considered as unsuitable for children. The OLFC (1999) guidelines specify the following for each classifiable element of the R category. It allows for sexual violence to be only implied and not detailed. Sexual activity may be realistically simulated but should not be actual sex. Nudity in a sexual context should not include obvious genital contact. There are virtually no restrictions on coarse language at the R level. The treatment of any adult themes with a high degree of intensity should not be exploitative. Finally, drug use should not be promoted or encouraged.

The other category that applies to video and then only in the Australian Capital Territory (ACT) and Northern Territory (NT) is the X category. It also applies to Internet content. The X classification only contains sexually explicit material and the OLFC guidelines (1999) specify the following for such material. It contains real depictions of actual sexual intercourse and other sexual activity between consenting adults, including mild fetishes. No depictions of sexual violence, sexualised violence or coercion, offensive fetishes or depictions that purposely debase or abuse for the enjoyment of viewers is permitted in this classification.

Films or videos that are RC cannot be publicly shown legally in Australia. This also applies to RC Internet content. The OLFC (1999) judges such material as offending against the standards of morality, decency and propriety generally accepted by reasonable adults. Such persons are defined officially as "possessing common sense and an open mind, and able to balance opinion with generally accepted community standards" (OLFC, 1999). RC material includes the promotion of crime and violence, depiction and promotion of paedophile activity and incest, and depiction of bestiality and other abhorrent phenomena.

The states of Victoria and Western Australia together with the Northern Territory were the first jurisdictions to regulate online services in 1996 (EFA, 2000). South Australia has now passed similar legislation (EFA, 2002). This was part of a national framework agreed upon in 1995 by the Standing Committee of Attorneys General (SCAG). Under such legislation the provision of "prohibited" or "potentially prohibited" material to minors is criminalized subject to two defences. First, the accused person did not know that the recipient was a minor and second, that the person took "reasonable steps" to avoid transmitting material to a minor. The states of New South Wales, Queensland and Tasmania together with the Australian Capital Territory are yet to pass similar legislation.

The federal government does not have the power to regulate publications, film, video or computer games, as this power is vested with the states and territories. They accept the OLFC classification codes although there are differences in use between them. The federal government has jurisdiction over television and broadcasting under its telecommunications powers and it has legislated on Internet content regulations under these powers. However, the federal government cannot prosecute content providers except for offences like child pornography, transmission of which is an offence under the Federal Crimes Act. Under the Broadcasting Services Act 1999 the Commonwealth can enforce take-down orders on content hosts. However, only the states and territories have the constitutional power to prosecute "offensive" content.

The federal legislation to regulate the Internet came into force on 1 January 2000 (COA, 1999) and is described as the Co-Regulatory Scheme for Internet Regulation. One of its objectives is to protect children from exposure to Internet content that is unsuitable for them. It is a complaints-driven system and the ABA can require an Internet Service Provider (ISP) to take down X rated and RC material hosted onshore and to take all reasonable steps to prevent access to X rated or RC material hosted offshore. The ABA must be satisfied that restricted arrangements, for example credit card access, are in place for material classified R originating onshore. No regulation exists for R material originating from offshore. The legislation expressly specifies time frames for the take-down process and penalties for non-compliance. ISPs are mandated to ensure that parents are aware of recommended filter technology products. The relevant Minister (Alston) reports every six months on the Scheme and examples are COA (2001, 2002).

# Key Questions

There are a number of key questions relevant to consideration of protection strategies for the protection of minors from harmful Internet content. These are addressed in this section.

## Who are children requiring protection?

Who are the children requiring protection from harmful Internet content? Are these children a homogeneous group that requires the same kind and level of protection? In Australian law a child is someone under the age of 18 years. However, it is obvious that when judgments are made about the need for, and level of protection, they should take into account the large disparity between, say a child of four years, and one of 17. Again, two children of the same age may require different kinds and levels of protection given differences in their level of emotional maturity and education, for instance.

In Australia this is recognized to some extent in the laws relating to regulation of the offline and online media. Films, for instance, may be classified G (General)—suitable for all ages, PG (Parental Guidance recommended)—for children under 15 years, M15+ (Mature Audiences)—for mature audiences 15 years and over and MA15+ (Mature Accompanied)—children under the age of 15 must be accompanied by a parent or adult guardian. Obviously R and X rated material, together with RC material, is forbidden for all minors to access and for adults, including parents, to knowingly provide access to.

## Why do children need protection?

Most would agree that children need protection from certain Internet content because consumption of it may cause harm to them. This is reflected in the parliamentary debates of the members from all parties (SH and HH) and parliamentary hearings by spokespersons from a diverse range of organisations (SSC). Certain Internet content may cause emotional trauma, for instance. Implicit is an acceptance that a person who has reached his or her late teens (18+ = adult) is able to exercise a mature choice as to what content to consume compared to someone younger (= child). Also assumed is that if an adult consumes potentially harmful content he or she is better able to minimize any harm that might be experienced. Without protection there is a greater likelihood that children may be accidentally be exposed to harmful content or that children may wilfully expose themselves to harmful content. The former is of most concern for younger children and the latter for older children

## Who decides what Internet content is harmful?

Again, most members of parliament in the debates (SH 5207; 4742) and spokespersons in the hearings (SSC128ff; 163-176; 256) argue that parents or guardians have the primary responsibility to regulate their child's access to Internet content and that a proper role for government is to empower parents with education and technology. Relevant industry and professional bodies also have a role in assisting parents with responsible behaviour guided by codes of conduct. The Australian government describes its regulatory framework as "co-regulatory" to emphasise it is in partnership with the relevant Internet industry and professional bodies.

The National Classification Code requires that classification decisions are to give effect, as far as possible, to a number of principles. Of relevance here is that minors should be protected from unsuitable material likely to harm or disturb them. This is officially defined (OLFC, 1999) as "material that is not appropriate to viewers under 18 years because of its ability to harm (cause development damage) or disturb (cause emotional trauma)". How then is it decided what is harmful or disturbing? The relevant "test" used in Australia is the standards of morality, decency and proprietary accepted by a reasonable adult.

To illustrate, the MA15+ Mature Accompanied material is considered likely to be harmful or disturbing to viewers under 15 years according to the OLFC. These guidelines state that:

- Generally depictions of violence should not have a high impact. Depictions with a high impact should be infrequent and should not be prolonged or gratuitous.
- Sexual activity may be implied.
- Coarse language that is very strong, aggressive or detailed should not be gratuitous.

- The treatment of (adult) themes with a high degree of intensity should be discreet.
- Drug use may be shown but should not be promoted or encouraged.

Some main differences between MA15+ and R (restricted to adults 18 years and over) is that in the latter strong depictions of realistic violence may be shown and sexual activity may be realistically simulated. There are no restrictions on coarse language, adult themes with a very high degree of intensity should not be exploitative and drug use may be shown but not gratuitously detailed. Thus, a minor may view a film classified MA15+ only when accompanied by a parent or guardian. A minor is not permitted to view a film or video classified R even if there exists parental permission and a willingness to accompany the child to view the film or video.

## What protection strategies are available?

There are two main protection strategies that may be employed to protect children from harmful Internet content. First is the strategy employed by parents/guardians and by extension, teachers. It is one primarily of effective supervision of what is accessed by the minor or what material the minor is exposed to. Such a strategy may employ filter software to block access to material considered harmful and education about how to deal with exposure to potentially harmful and disturbing material. Second is government legislation that identifies what is illegal and what is potentially harmful or disturbing to children. It then legislates to make certain content illegal or restricts access based on the age of the minor. Ultimately the issue is what is the "right" balance between the two broad strategies.

Classification of content can provide valuable consumer advice to both child and parent about material that may be harmful. The government can also assist with public awareness and educational programmes and ensure filter technology is accessible to parents. The federal government's community advisory panel, called Net Alert, has these as part of their mandate. The Internet industry is seen to assist with codes of conduct and making accessible cheap filter software available to ISP subscribers, for instance.

# Controversies, Problems and Solutions

## Regulation and the Classification Code

The National Classification Code (and its state and territory variants) provides helpful consumer advice to parents and guardians about what minors should access. However, its primary role is censorship of what both adults and children can access. The regular meeting of State and Commonwealth Attorneys-General (SCAG) is widely known as the meeting of censorship ministers. Parental disregard of some "consumer advice," for instance giving parental permission for a child to attend a film classified MA15+ unaccompanied by an adult, is an offence. Under some states' laws parents who knowingly provide access to "prohibited" material to their child commit a criminal offence. If parents inadvertently provide access to X rated material to their child then the onus of proof is on the parent to prove that reasonable steps were taken to avoid access. Some states' laws mean that parents must make judgements about material not yet classified. They must judge it to be "potentially prohibited" because when classified this may turn out to be the case and so make them liable to be charged. Parents must make judgements that the "experts" on the Classification Board find difficult and who are often not unanimous in their classification decisions (EFA, 1999).

## Regulation and the "Reasonable Adult"

Under the Classification (Publications, Films and Computer Games) Act 1995, the OLFC Board must make classification decisions based on the Code. It is to give effect as far as possible to the following principles:

(a)   adults should be able to read, hear and see what they want,

(b)   minors should be protected from material likely to harm or disturb them,

(c)   everyone should be protected from exposure to unsolicited material that they find offensive, and

(d)   the need to take account of community concerns about depictions that condone or incite violence, particularly sexual violence and the portrayal of persons in a demeaning manner (OLFC, 1999).

Supporters of regulation, like Attorney General Williams (1991), claim that the free speech rights of adults are provided for whilst at the same time children are afforded protection, together with adults, from offensive and other material of concern to the community. They believe that the "correct" balance is struck between adult freedoms and children's protection (SSC 4ff; 128ff). However, one conservative politician at the hearings, after learning that no parents had expressed concern to Internet access in libraries exclaimed with some honesty, "we have full censorship in this country, we always have. Are you saying that libraries should be totally free from that" (SCC, 196).

In classifying content for censorship and consumer advice, the test of the "reasonable adult" and the concept of a "community standard" are paramount. Australian society is composed of many communities of widely differing standards in relation to sexuality, gender relations and adult themes. One of Australia's greatest assets is a social structure that is highly pluralistic. In reality, the "test" results, at best, in the majority norm (conservative) being forced on all communities (Weckert, 1998; Whittle, 1998). Censorship by government and the primary right of parents to decide what their children access on the Internet is a relationship of high tension. Invariably, government regulation constrains parental decisions about what their children may access on the Internet. This occurs at the expense of empowerment of parents and at the expense of adult free speech (HH, 6914).

## Regulation and Community Concerns

Supporters of Internet censorship legislation, as with other media like film, video and computer games, argue it represents a response to community concern about harmful effects of certain Internet content (HH 6907; SH 5218). This is especially so for the perceived concerns about the harmful effects on children and especially the alleged harmful effects of sexually explicit content that is invariable described as "pornography". It is described as a major social problem (SH, 5218) and for some a belief that over 60% of all Internet material is pornographic (SSC 49; 73). As one member of the House put it, this material is "…a litany of filth and cyber-septic that people are up against" (HH, 6917) or as a spokesperson at the Hearings put it, "because industry has not done anything about it our homes are invaded by groups—including pimps and prostitutes—pushing their porn, causing parents a great deal of concern" (SCC, 5215).

Thus, the legislative responses by the federal and state, and territory governments are seen as a proper response by government to meet this perceived community concern. The relevant Federal Minister asserts the legislation makes for a safer environment for Australian children (Alston, 2001, 2000) and claims there has been little or no criticism of it from the community at large (SH, 5218).

What empirical evidence exists suggests that harm to minors through exposure to pornography is not a major community concern where the Internet is concerned. This is not to say that most parents are not concerned about what their children read, hear and see, because they are. However, when asked in surveys and questionnaires, the issues of security, privacy and cost regarding the Internet are most important (SSC, 215-216; 260ff; 297-299).

## Regulation and the Nature of the Internet

In seeking to protect children, government regulators betray a lack of understanding about the nature of the Internet. The federal government treats it like a film for content classification purposes but gives regulatory responsibility to the ABA, whose regulatory expertise is in broadcasting. Indeed its long history in regulation probably persuaded the federal government to choose it as the regulatory authority (SCC, 133). Indeed the ABA at the hearings confirmed that they could administer and enforce the legislation (SSC, 31; 87-88). In the period right up to the tabling of the Online Services Bill there were many in the government who believed that in the Australian context that X rated content deemed harmful to minors could be blocked at the ISP or backbone provider level or that ISPs could ensure that R rated material was subject to adult verification (EFA, 2000). In other words, a regulatory system similar to that attempted in a country like China or Saudi Arabia could be implemented. When the legislation was finally passed there was belated recognition that little could be done to regulate overseas content, which comprises over 90% of "harmful" material accessible to minors (SH, 5218).

Aspects of the Internet may be likened to a diverse collection of offline media, but it is these and more (SCC 47ff; 67; 119; 132ff; 208ff; SH 5205ff). Indeed in the future it is likely that the Internet will continue to evolve and will comprise more and more complex and convergent communications. Application of a regulatory framework designed for something else is likely to be ineffective or unenforceable in protection of minors. This appears to be the case for Australia as it was for the United States in relation to the Communications Decency Act (Sandy, 2002, 2001b).

## Regulation and Harm to Minors

Supporters of legislating to censor the Internet claim that its major justification is to protect children from the harmful effects of Internet content, which inevitably is equated with "pornography" (HH6916-6918; SH, 5215). Most proponents of government regulation accept as a given that sexually explicit material is harmful to minors. However, there is little empirical evidence to support a causal link between pornography and harm to minors (Sandy, 2001a). However, the difficulties in conducting experiments with children, especially young children, are acknowledged. In the case of adults there exists a great deal of empirical work undertaken over many years on the harmful effects of pornography. On balance it does not support a causal link between "consuming" pornography and "bad" behaviour (Sandy, 2001a).

Many argue that the best antidote against potential harm for children is to educate them about sexuality, violence and adult themes and that they be nurtured in a secure and loving environment (Carol, 1994; Mills, 1992; Small, 1990). The greater the ignorance of the child, the more likely the harm will be. This is the primary responsibility of parents, supported by government and school. Further, it is argued that censorship acts to promote certain content as "forbidden fruit," increasing its attraction to children. Censorship is thus counterproductive.

In the debates and hearings much was made of child pornography and the need for regulation of the Internet to combat it (SSC 215-216; 262-273). However, those opposed to the Online Services Bill pointed out that child pornography was already a criminal offence under the Crimes Act and therefore "special" Internet legislation was not needed (SSC 48-49; 209). Consideration should be given as to whether a distinction should be made between those who produce the images and those who consume them. Most would agree that the involvement of children in the production of sexually explicit images is non-consensual simply because the children are minors and in law are unable to consent. Therefore it should be an illegal adult activity. What is less clear is whether it should be illegal for adults to consume these images. Less clear is whether access to images of "adults" masquerading as children should be illegal. In some Australian jurisdictions both are criminal offences.

# Regulation and its Effectiveness

Supporters of federal regulation and its state and territory counterparts argue that regulation has been effective in protecting minors from harmful Internet content and supporting parents through education and technology. In media releases, Minister Alston (2000; 2001) claims the Internet is a safer place for children, indeed for all Australians. The Minister refers to instructions to ISPs to take down illegal and highly offensive material from the Internet as evidence.

Opponents of the federal and state/territory legislation argue that it has not made the Internet a safer place for minors. This is because 90%+ of material that is potentially harmful to children is hosted offshore and beyond Australian jurisdiction (SSC, 63ff; HH, 7970; SH, 5136). Unless Australia is willing to adopt draconian controls over the Internet, like China or Saudi Arabia for instance, this remains the case. The federal government has been unwilling to adopt such an arrangement and has recommended that families be provided with filter software and has instructed the Commonwealth Scientific and Industrial Organisation (CSIRO) to evaluate alternative filter products (CSIRO, 2001). The ABA now recommends provision of specified filter products despite a recognition that they vary considerably in their effectiveness in protecting minors from access to harmful material.

In addition it is claimed (EFA, 2001) that the ABA has overstated the amount of material, mostly child pornography, removed from the Internet and cites a lack of police prosecutions in regard to child pornography material. The Australian Institute, a strong supporter of Internet regulation, asserts, "not only the regulation of pornography on the Internet is manifestly failing but the regulatory authorities themselves appear to have lost sight of their functions" (Flood & Hamilton, 2003). It claims the ABA seems more concerned to promote the use of the Internet than to protect children from its dangers. The EFA (2001) and others (SSC, 256; HH, 6911-6914) suggest that Minister Alston engages in misleading political rhetoric that grossly overstates the effectiveness of the legislation. They point to the risk that this lulls Australian parents into a false sense of security by thinking their children are "safe".

# Children Requiring Protection

Most would agree that allowance must be made for age differences in regard to the group defined as children in the law. Flexibility is desirable. A person over

18 years may be "harmed" by certain Internet content but a child may not be "harmed" by the same content. So long as the classification serves as consumer advice parents can decide about access. Flexibility disappears when the state makes it an offence for the parent to permit a child access to content the state has declared harmful. In a recent review (Brand, 2002) of the classification guidelines for films and computer games it is recommended that a more comprehensive age based system be adopted. This makes a further distinction between children eight years and over and 12 years and over. This provides a modicum of extra flexibility.

## Decision on Harmful Content

Most agree that parents have primary responsibility for what their children access on the Internet. However, supporters of government legislation argue there is a need to assist parents in this regard. First, is a concern that computer literate children will deceive their computer illiterate parents and access content potentially harmful to them (SSC, 51; SH, 5219; HH, 7977). Therefore government should make decisions on their behalf. A problem is that it is made for all parents, not just those who are computer illiterate, and is given whether they want it or not. Second is a concern that parents do not undertake their responsibility to properly supervise their children and so legislation is required to deal with these irresponsible parents (SSC, 51; SH, 5212; 5219). This is akin to protecting children from their own parents. However, there is community acceptance on the "right" of government to override parental rights in certain circumstances, in which a child is abused or neglected by its parents, for instance. This right of the state is relative, as its "rightness" can change over time. For decades in Australia it was considered a "right" for the Government to take away mixed blood children from their Aboriginal parents and foster them with white families or place them in institutions. Today this policy would not command widespread support.

## Protection Strategies

Government in Australia exhibits trust in the effectiveness of filters for use at home, schools and in libraries. Under the federal legislation, for instance, each ISP must make available cheap filters for service subscribers. However, there is much evidence that filters are notoriously ineffective (CSIRO, 2001). They fail to block what they intend to block and block what is not intended to be blocked. The issue here is that excessive claims by government about filters may lull parents into a false sense of security, as has been previously mentioned. The misleading rhetoric has the effect of being counterproductive. The adoption of a domain, like .adu, that houses legal sexually explicit material in the United States and Europe would assist Australian regulators to isolate legal material suitable for adults from access by children. It could be mandated that such material be placed on this domain. This is likely to improve the effectiveness of filters. The danger is that it is easier to censor what adults access and so violate their free speech rights.

Developments like instant messaging and file sharing services may prove more difficult to regulate for the protection of children than the WWW has. Recently in Australia the Attorney Generals have met to discuss the regulation of mobile phone cameras. There is a concern that the technology permits photos of children (and adults) being taken and placed on the Internet. Local government in some areas of Australia have banned such technology from municipal swimming pools.

## Summary and Conclusion

In a society like Australia there is agreement by most that adults should be able to read, hear and see what they want. However, this is within a consensual boundary, one that can shift back and forth over time. Some speech is deemed illegal or restricted to certain groups or circumstances. The test of what should be censored in Australia is that of the reasonable adult.

Australia has a long history of censorship of offline speech. This is justified on the grounds that some speech is harmful to adults and/or minors. Recently, the regulatory framework was extended to the Internet and justified on the same grounds.

This chapter has addressed how to best protect minors from harmful Internet content whilst preserving the free speech of adults. The regulatory framework has not been successful in protecting minors from harmful Internet content. The main reason is that over 90% of material described as pornographic is from offshore and largely beyond Australian jurisdiction.

This emphasis on a regulatory approach, notwithstanding claims to the contrary, means that empowerment and education of families has been neglected. The reliance on ineffective filter software has compounded the problem. Indeed, families may be lulled into a false sense of security that the framework provides a high level of protection to minors from harmful material.

The regulatory approach has not preserved the free speech of adults. Supporters claim that what is censored online should be the same as what is censored offline. Even if we accept the government censorship of much of the offline material as justified, the Internet is treated differently (more harshly) than offline media. Onshore generated X rated material, which is sexually explicit but nonviolent, is subject to take-down notices. ISPs are required to take all reasonable steps to prevent access to offshore generated X rated material. As an aside this means that extremely violent material, which some claim to be harmful to adults (and minors) can still be accessed.

In summary, the Australian Internet regulatory framework has had limited success in protecting minors from harmful Internet content and it does compromise the free speech rights of adults.

# References

Alston, R. (2000). *A safer Internet for all Australians*. Media Release, 19 May.

Alston, R. (2001). *Australia's safer Internet*. Media Release, 13 February.

Bland, J. (2002) *A review of the classification guidelines for films and computer games: Assessment of public submissions on the discussion paper and draft revised guidelines*. Bond University Centre for New Media Research and Education, 11 February.

Carol, A. (1994). *Censorship won't reduce crime* [Online]. Available: http://www.libertarian.org/LA/censcrim.html

Commonwealth of Australia. (1999a). *Senate Hansard: Broadcasting services amendment (Online Services) Bill 1999 – Second Reading*. Canberra.

Commonwealth of Australia. (1999b). *Official committee Hansard: Senate select committee on information technologies – Broadcasting services amendment (Online Services) Bill 1999*. Canberra.

Commonwealth of Australia. (1999c). *House Hansard: Broadcasting services amendment (Online Services) Bill 1999 – Second Reading.* Canberra.

Commonwealth of Australia. (1999d). *Broadcasting services amendment (Online Services) Act 1999.*

Commonwealth of Australia. (2001). *Sixth-month report on co-regulatory scheme for Internet content regulation July to December 2000.* Department of Communications, Information Technology and the Arts, April.

Commonwealth of Australia. (2002, February). *Sixth-month report on co-regulatory scheme for Internet content regulation January to June 2001.* Department of Communications, Information Technology and the Arts.

Commonwealth Scientific and Industrial Research Organisation. (2001, September). *Effectiveness of Internet filtering software.* Mathematical and Information Sciences.

Electronic Frontier Australia. (1999). *Censorship blinded by the smoke: The hidden agenda of the Online Services Bill 1999* [Online]. Available: http://rene.efa.org.au/liberty/blinded.html

Electronic Frontier Australia. (2000). *Internet regulation in Australia* [Online]. Available: http://www.efa.org.au/Issues/Censor/cens1.html

Electronic Frontier Australia. (2001). *Regulatory failure: Australia's Internet censorship regime* [Online]. Available: http://www.efa.org.au/Analysis/aba_analysis.html

Electronic Frontier Australia. (2002). *South Australian Internet Censorship Bill 2002* [Online]. Available: http://www.efa.org.au/Campaigns/sabill.html
Flood, M., & Hamilton, C. (2003, February). *Youth and pornography in Australia: Evidence on extent of exposure and likely effects.* Discussion Paper No.52. The Australian Institute.

Mills, J. (1992). Classroom conundrums: Sex education and censorship. In L. Segal & M. McIntosh (Eds.), *Sex exposed: Sexuality and the pornography debate.* Virago Press.

Office of Film and Literature Classification. (1999). *Guidelines for the classification of films and videotapes (Amendment No. 2.).*

Sandy, G. (2001a). The Online Services Bill: Theories and evidence of pornographic harm. In J. Weckert (ed.), *Conferences in research and practice in information technology: Computer ethics, 1,* 46-55.

Sandy, G. (2001b). Public and private censorship of the Net – Stories e-commerce professionals need to know about. In A. Wenn (Ed.), *Proceedings of the SSECP Conference: Skill sets for the e-commerce professional*. Melbourne, Australia.

Sandy, G. (2002). The effectiveness of the co-regulatory scheme for Internet content regulation. In M. Warren & J. Barlow (Eds.), *Proceedings of the Third Australian Institute of Computer Conference*. School of Information Technology, Deakin University, Australia.

Small, F. (1990) Pornography and censorship. In M. Kimmel. (Ed.), *Men confront pornography*. Crown Publishers Inc.

Weckert, J. (1998). Offence on the Internet. *Res Publica, 7*(1).

Whittle, R. (1998). *Executive summary of the Internet content debate: A comparison table between broadcasting and Internet communications and a critique of Peter Webb's speech* [Online]. Available: http://www.ozemail.com.au/~firstpr/contreg/bigpic.htm

Williams, D. (1991). *Censorship and human rights: Finding a legal balance*. B'nai B'rith Anti Defamation Commission.

Chapter VIII

# Mobile Communities and the "Generation that Beeps and Hums"

Marian Quigley
Monash University, Australia

## Abstract

*The rapid appropriation of mobile phone technology by young people is occurring at the same time as critics are debating the so-called demise of community, purportedly as a result of our increasingly technologised and globalised society. Opposing theorists, however, argue that the notion of community is itself nebulous and that it represents an imagined ideal rather than a vanishing reality. Thus, they argue, it follows that debates about the greater authenticity of "real," face-to-face communities over "virtual communities"—those centred on technological rather than geographical links —are based on a false premise. This chapter argues that young people today are utilising mobile phones—sometimes in combination with the Internet—to establish and maintain social networks combining both their geographically present and absent peers. These networks are mobile, heavily reliant on technology and are comprised of a mix of "real" and "virtual" communication. They are also characterised by a sense of belonging to a group—a concept integral to the notion of community.*

# Introduction

*The adoption of the mobile phone by teens is a new area that goes beyond our experience with the traditional telephone. This is uncharted territory as no generation of teens has had access to this type of technology.* (Ling & Helmerson, 2000, pp. 8-9)

*[T]echnology is always, in a full sense, social.* (Williams, 1981, p. 227).

As Alexandra Weilenman and Catrine Larsson note, researchers have only recently become interested in the "social and interactional aspects" of the uses of mobile phones (2002, p. 92). A number of studies focusing primarily on young people's use of mobile phones are also beginning to emerge along with the realisation of the high level of adoption of this communications technology by this demographic, particularly in developed nations. Rich Ling has undertaken a number of individual and collaborative studies of teenagers and their utilisation of mobile phones in Norway. Katz and Aakhus's *Perpetual Contact: Mobile Communication, Private Talk, Public Performance* (2002) also includes studies of Finnish and Norwegian teenagers and mobile phone use. Howard Rheingold's recent book, *Smart Mobs: The Next Social Revolution* (2002) examines the utilisation of mobile phone technology for both social and political aims by young people in a number of countries, including Scandinavia, the Philippines and Japan. In Australia, a smaller study by Carroll et al. (2001) examines the appropriation of information and communication technologies (ICTs), including mobile phones by young people, whilst social researcher Hugh Mackay (2002) observes the ways in which young Australians are using mobile phones alone or in tandem with the Internet to establish new forms of community. All of these studies demonstrate similar findings regarding the popularity of mobile phones amongst young people, the underlying reasons for this popularity and the ways in which mobile phones are utilised.

As Raymond Williams argues, technology is a *relationship* that is "necessarily in complex and variable connection with other social relations and institutions" (1981, p. 227). Similarly, Howard Rheingold notes that the "'killer apps' of tomorrow's mobile infocom industry won't be hardware devices or software programs but social practices" (2002, p.xii). A number of researchers have found that ICTs are used in ways that reinforce existing social relationships at the same time as they appear to be transforming our lives. The ways in which young people use mobile phones demonstrate that technology is a shaped, as well as a shaping, phenomenon. For them it is often a shared resource rather than merely a privatised medium of communication. Moreover, their preference for Short Message Service (SMS) text messaging is a phenomenon that was unforeseen by manufacturers.

Young people are utilising mobile phone technology to establish and maintain peer-based social networks at a time when theorists are debating technology's role in the apparent loss of community. Meanwhile, others argue that the concept of community is imagined rather than actual, therefore rendering meaningless the debate concerning the greater authenticity of real as opposed to virtual communities.

# The Loss of Community

> "Community' is nowadays another name for paradise lost – but one to which we dearly hope to return ...." (Bauman, 2001, p. 3)
>
> "...all communities larger than primordial villages of face-to-face contact (and perhaps even these) are imagined." (Anderson, 1991, p. 6)

While it appears that the longing for "community" is universal among human-kind, the term *community* itself lacks a firm definition—although it has enduringly positive connotations. In a recent article, Maria Bakardjieva, drawing on Raymond Williams' etymology of the term, notes "its interpretative flexibility and hence its socially constructed character [and that there] is no consensually accepted definition of its meaning" (Bakardjieva, 2003, p. 292). Williams himself describes community as "the warmly persuasive word to describe an existing set of relationships; or the warmly persuasive word to describe an alternative set of relationships" that "seems never to be used unfavourably and never to be given any positive opposing or distinguishing term" (Williams, 1985, p. 76).

Although one is located in the public, the other in the private realm, *community* can be seen to share similar qualities to those of *home:* that "firm position which we know, to which we are accustomed, where we feel safe, and where our emotional relationships are at their most intense" (Heller, 1984, p. 239). But, as media theorist Roger Silverstone concedes, the concept of community tends to remain bound to physical place:

> *... it is difficult to think of community without location, without a sense of the continuities of social life which are grounded, literally, in place. Community, then, is a version of home. But it is public not private. It is to be sought and sometimes found in the space between the household and the family and the wider society.* (Silverstone, 1999, p. 97)

Commenting on the public perception of the loss of close-knit communities, the renowned sociologist Zygmunt Bauman also links the notion of community to place. He notes the erosion of the "epistemological foundation of the experience of *community*": of "the steady and solidly dug-in orientation points"— such as the 'friendly' bank and corner store—"which suggested a social setting that was more durable, more secure and more reliable than the timespan of an individual life". He argues that "no aggregate of human beings is experienced as 'community' unless it is 'closely knit' out of biographies shared through a long history and an even longer life expectation of frequent and intense interaction". Bauman claims that it is due to the absence of such experience that commentators today often report on the "demise" of community (2001, pp. 47-48).

## The Role of ICTs

Critics often associate the loss of community with the increased technologisation and resulting privatisation of society. The erosion of traditional forms and places of family and community that we are now witnessing began with the Industrial Revolution that was accompanied by what Raymond Williams describes as the phenomenon of "mobile privatisation" (1981, p. 228). As a result of the development of an industrial economy, the population was dispersed; places of work and home were separated and community members were isolated from one another in privatised forms of living. However, earlier communication and information technologies (ICTs) such as the telephone, radio and newspaper enabled the maintenance of links with geographically dispersed communities. More recently, the television, computer and mobile phone have further eroded the boundaries between work and home and, at the same time, undermined the significance of place, time and space.

In the past, physical places such as the home and the factory defined social roles and status. A change in social situation resulted from movement from one place to another across space and time. Doorways, which enable inclusion or exclusion, marked social roles or place. As a consequence of their ability to permeate walls and travel great distances almost instantaneously, ICTS "destroy the specialness of space and time" and bypass "the social rite of 'passage'." They also destroy the specialness of place—a uniqueness attributed by the activities associated with it—by turning public spaces into private ones and private spaces into public ones (Meyrowitz, 1997, pp. 43-50). Furthermore, it is as a result of ICTs that our world has become "for the first time in modern history … relatively placeless". "Wherever one is now—at home, at work, or in a car—one may be in touch and tuned-in" (Meyrowitz, 1999, p. 100).

Large-scale societal change accompanied by rapid and increasingly mobile technological development in the late twentieth and early twenty-first centuries has therefore resulted in the undermining of traditional familial divisions within the home along with the dissolution of the boundaries between the private home and the wider community. This can cause feelings of insecurity and/or anxiety amongst both adults and young people who still feel the need to belong to a secure, familiar place within which they know their social place. We all want to belong to a community, but as a result of fragmenting culture, fractured experience and social and geographical mobility, we have come to lack certainty as to what it is. Young people today are growing up within homes of mixed or single parent families; within societies that lack a shared value system such as that previously offered by religious belief; and in an era in which so-called "multiculturalism" and "globalisation" compete with localism/regionalism. Furthermore, where once so-called "mass communication" purportedly held a "passive" audience together in the home/community, the proliferation and increased individualisation of ICTs have fragmented the now more obviously "interactive" audiences. The availability of ICTs also means that it is now virtually impossible to shield the young from graphic depictions of global terrorism such as the events of September 11: events that subverted notions of durability and security on a grand scale and that have most likely heightened the need for Bauman's "steady and solidly dug-in orientation points".

Alongside the technology detractors (see, for example, Neil Postman's *Amusing Ourselves to Death* and *The Disappearance of Childhood*) who have blamed ICTs for the demise of communities (along with a range of other societal ills including a decline in morality, increased violence, consumerism, loss of local cultures, etc.), theorists such as Nicholas Negroponte and Howard Rheingold have previously lauded new technologies—particularly the computer—as harbingers of new forms of virtual electronic communities. Others such as Postman, meanwhile, cling to an idealized notion of real, non-technologically mediated community. This debate has resulted in the establishment of a dichotomous division between virtual communities and idealised real, authentic communities grounded in immediacy and locality (Bakardjieva, 2003, p. 293). In the revised edition of his book, *Virtual Communities* (2000), however, Rheingold himself concedes the utopian nature of his earlier vision of computer-mediated communication. He now believes "online social networks"—a term coined by the sociologist Barry Wellman—to be more appropriate than "virtual communities".

Wellman deconstructs the value-laden term *community,* questioning the likelihood of its existence—at least in its idealized form—arguing that, in any case, it is not applicable to modern Western societies. He debunks the "standard pastoralist ideal of in-person, village-like community" in which individual community members provide broad support to one another, stating that,

> *It is not clear if such a broadly supportive situation has ever actually been the case—it might well be pure nostalgia—but contemporary communities in the western world are quite different. Most community ties are specialized and do not form densely knit clusters of relationships ... People do get all kinds of support from community members but they have to turn to different ones for different kinds of help."* (Wellman, 1992, in Rheingold, 2000, p. 363)

Wellman also distinguishes the notion of community from that of neighbourhood, defining community as "networks of interpersonal ties that provide sociability, support, information, a sense of belonging and social identity" (in Rheingold, 2000, p. 360). He therefore severs community from place.

> *Although people often view the world in terms of groups, they function in networks. In networked societies, boundaries are permeable, interactions are with diverse others, connections switch between multiple networks, and hierarchies can be flatter and recursive. The change from groups to networks*

*can be seen at many levels ... Communities are far-flung, loosely-bounded, sparsely-knit and fragmentary. Most people operate in multiple, thinly-connected, partial communities as they deal with networks of kin, neighbours, friends, workmates and organizational ties. Rather than fitting into the same group as those around them, each person has his/her own 'personal community' ... [Wellman prophesies that] the person-not the place, household or workgroup-will become even more of an autonomous communication node ... people usually obtain support, companionship, information and a sense of belonging from those who do not live within the same neighbourhood or even within the same metropolitan area. People maintain these community ties through phoning, writing, driving, railroading, transiting and flying.... The person has become the portal.* (Wellman, 2001, in Rheingold, 2002, pp. 56-57)

Bakardjieva draws on the work of both Barry Wellman and Benedict Anderson, arguing that the dichotomy between virtual and real-life communities is false since the "majority of the so-called 'real-life' communities are in fact virtual in the sense that they are mediated or imagined" (Bakardjieva, 2003, p. 293). Bakardjieva prefers the term "virtual togetherness" to "virtual community" and questions the dichotomy between public and private (p. 292). She argues that the cultures enacted online cannot be divorced from "real" encounters as they have their roots in forms of life in the real world (p. 294). "The opposite of virtual togetherness (and community) is not 'real' or 'genuine' community, as the current theoretical debate suggests, but the isolated consumption of digitised goods and services within the realm of particularistic existence" (p. 294).

Although Bakardjieva mainly focuses on computer-mediated communication, her argument can also be applied to mobile telephony. However, whilst the author acknowledges the usefulness of Wellman's, Rheingold's and Bakardjieva's debates, she would like to reclaim the term "community," albeit as a reconstructed one divorced from place and characterised by a series of changing, loose and close communication ties. Teenagers' use of mobile phones involves communication with peers who are both geographically present and absent, comprising established social networks as well as the forging of new ones in what the author prefers to describe as "mobile youth communities". As a number of studies reveal (Kasesniemi & Rautiainen, 2002; Weilenmann & Larsson, 2002), the manner in which young people use the technology is often based on a sharing of communication messages and/or the device itself, rather than a practice of "isolated consumption". Young people have become the mobile portals acting to sustain communities within fragmented post-industrial societies.

# Mobile Phones and Young People

*Another issue that the results from this study raise is the way teenagers spend time in groups, and how this is shown in their use of the mobile phone.* (Weilenmann & Larsson, 2002, p. 105)

*The shift to a personalized, wireless world affords networked individualism, with each person switching between ties and networks. People remain connected, but as individuals rather than being rooted in the home bases of work unit and household. Individuals switch rapidly between their social networks. Each person separately operates his networks to obtain information, collaboration, orders, support, sociability, and a sense of belonging.* (Wellman in Rheingold, 2002, p. 195)

Ownership of mobile phones has become ubiquitous amongst teenagers, who regard them as almost a necessity in order to maintain social links with their peers. Mobile phones are also closely associated with a sense of self-identity. Additional factors such as fashion, convenience, utility and cost are important considerations for young people with regard to mobile phone usage. The fashion industry, in recognition of the ubiquitous nature of teenagers' ownership of mobile phones now incorporates special pockets to hold them within clothing and accessories. The device needs to be durable enough to withstand its carriage in school bags, pockets of trousers or school uniforms and not so small that it becomes mislaid or that it is too difficult to type text messages on its keys. Prepaid phone cards are often found to be the best way to manage costs (Carroll et al., 2001, p. 100). One of the most important factors relating to mobile phone ownership amongst teenagers is its ability to offer independence.

## The Role of Adolescence in Contemporary Society

The division between parent and child is a socially constructed one that is reinforced and maintained by social institutions such as the family and school and that, in encouraging behaviour "appropriate" to the child, tend to produce that behaviour. Childhood is "historically, culturally and socially variable" and is defined by its relation to "adulthood," another unfixed term (Buckingham, 2000, pp. 6-7). Adolescence, then, also poses problems in definition.

The period of adolescence is unique within industrial and post-industrial societies. As opposed to pre-industrial societies, the child's future professions and life experiences are unlikely to mirror those of the child's parents.

Consequently, the adolescent now plays an active role in the formation of his/her transition from childhood to adolescence and particularly from adolescence to adulthood. At the same time, the expansion of the educational system that resulted from the need for highly skilled labour has led to the school's increasing role in socialisation and an expanded period of youth. "In addition, however, the age grading of the school system means that one's same-aged peer group takes a central role in the youth's activities, their sense of identity, consumption patterns and in their orientation" (Ling & Helmerson, 2000, p. 3). The peer group plays a central role in the life of the adolescent who is in a transitional period between living under his/her parents' roof and rules to moving out of the family home and living an independent existence. Drawing on a number of sociological studies, Ling and Helmerson note that while "the pre-adolescent's relationship to adults provides a sense of ordered reality, the peer group provides one with the sense that they can modify social interactions and thus these relationships provide mutual meaning" (2000, p. 4).

## The Mobile Phone as a Marker of Maturity and Independence

As Ling and Helmersen point out, the child is acquainted with the traditional telephone from infancy and from the age of seven years, has generally gained the requisite motor skills along with parental permission to use the device. However, the ability to manage a conversation depends on greater cognitive and social development, including the ability to put oneself into the position of the other speaker and to understand it. This is generally achieved between the ages of 12-14 years. In addition to this, the ownership and use of a mobile phone is dependent upon the skills to manage both a budget and a social network (Ling & Helmersen, 2000, pp. 6-7).

As Rich Ling notes in an earlier study, "discussions about teens' ownership and use of technology are a moment in the social definition of their maturation" alongside other markers such as the first use of make-up or alcohol (1999, p. 2). The addition of a "mobile communication component" to the teenagers' media use

> provides the potential for coordination and interaction within the peer group
> in a way that was impossible up to now ... the TV allows the child to view the
> entire world.  The mobile telephone now allows them to interact without the

*traditional filtering of parents. Thus, the peer group can and is being orga-nized in new, more dynamic ways.* (Ling, 1999, p. 3)

Communication between peers is essential to the maintenance of the peer group and the individual's identity. The mobile phone—a consumer item that is assigned significant cultural capital by today's adolescents—enables ongoing communication between family members and peers both within and outside the parental home and is thus emblematic of the transition from childhood to adulthood.

## Empowerment and Socialisation

*Teens use the messages to test their limits and step outside the role of a child ... Text messaging is a way to share relationships.* (Kasasniemi & Rautianen, pp. 180-182)

Mobile phones function as a tool of empowerment and socialisation for young people, as a number of recent international studies demonstrate. A national report commissioned by the Australian Commonwealth Consumer Affairs Advisory Council (2002), for example, found that mobile phones are probably the most important product for young people. "Respondents said mobile phones symbolised freedom, growing up, excitement and having fun and were 'must haves' for teenagers wanting to keep up and achieve social acceptance" *(http://parlsec.treasurer.gov.au)*. In addition, a South Australian poll (2002) found that almost 70% of youths send more text messages than mobile phone calls. SMS was regarded as a useful, cheaper and more discreet tool for approaching members of the opposite sex (Kendall, 2002, p. 11). A Melbourne University study found that young people used mobile phones at home, at school and university to maintain existing friendship groups; to establish and reinforce individual and group identity; to exercise power and to "achieve a sense of cohesion by dealing with the fragmentation of their lives" (Carroll et al., 2001, pp. 98-99). The sense of belonging to a group was reinforced as a result of frequent calls between current and former friends.

In accordance with a number of overseas findings, Carroll et al.'s study revealed that the mobile phone seems to be regarded by young people today as a necessity for the maintenance of a social life. Those without mobile phones experienced difficulty in maintaining their social networks. Additionally, mobile phones allow for last-minute arrangements or changes to social plans, as one respondent explained:

> *Before I had a mobile, I used to make strict plans meeting friends and family ... Now, with mobile phones, everyone's got one. Within 15 minutes of meeting up it's 'I'll see you at so-and-so'.* (Carroll et al., p. 100)

Interestingly, Carroll et al. found that amongst some of the respondents, the choice of an ICT was related to the relative intimacy of the relationship. SMS and the mobile phone were used for the most intimate relationships; e-mail was chosen for communication with absent family and friends or with friends in previous social networks who were not often in face-to-face contact; and Internet chat was used for less intimate acquaintances and friends made through chat sessions (Carroll et al., 2001, p. 99).

The study also found that young people use ICTs to exercise power. Mobile phones enable the filtering of calls so that children can choose not to answer their parents. Phone calls can also be made from the child's bedroom unbeknown to the parents and SMS allows communication between students within classrooms without teachers' knowledge. The broadcasting of an SMS message to a group of friends prevents arguments about social arrangements due to the lack of dialogue (Carroll et al., 2001, p. 99). ICTs also make an important contribution to the development of a strong sense of self-identity and membership of a group which frequent calling of friends reinforces. In the words of one interviewee: "[The mobile phone] gives you an identity: this is who I am, this is *my* number" (Carroll et al., 2001, p. 98). SMS messages play a particularly important role in the maintenance of group identity.

Carroll et al. concluded that ICTs have "shaped the young peoples' lives and facilitated new ways of interacting" whilst themselves being adapted to the users' needs. They predict that "the use of ICTs will continue while they enable young people to establish a sense of identity and belonging, gain power over important figures in their lives and develop cohesion in an increasingly fragmented world" (Carroll et al., 2001, p. 99).

# Technology Appropriation by the "Generation that Beeps and Hums"

*The media habits and cultural preferences of youth challenge the foundations of norms ... At the same time, it is a matter of a struggle over cultural capital. By being curious about the new the young acquire skills that adults lack: young people know more about comics, films, video machines and computers than most adults do. The older generation sometimes experiences this as threatening and in response try to put young people in their place.* (Boethius, 1995, pp. 48-49)

Young people are often regarded as being at the forefront of technological expertise. Variously described as "the electronic generation"; "the avant-garde of consumption"; "Generation txt"; and so on, they tend to master new technology and incorporate it into their lives quickly, and, in doing so, often discover new ways of utilising it. In their Australian study, Carroll et al. found that young people tend to make "technology-driven rather than task-driven decisions," working from the premise of how the technology can be incorporated into their lives rather than how it can help them complete a task (2001, p. 100). Adolescents' use of mobile phones for text messaging is one example of how (young) people turn devices to their own creative uses (Strathern, 1992, p. x).

Young people also have been found to use the mobile phone in communal as well as privatised ways that involve both co-present and absent peers. Weilenmann and Larsson discovered that Swedish teenagers used the mobile phone as "a tool for local social interaction, rather than merely as a device for communication with dislocated others" (2002, pp. 92-93). Rather than using the device primarily for private communication, mobile phones and their messages were often shared between co-present friends. Teenagers took turns in speaking to the caller and text messages were either read aloud or the screen displayed to other members of the group. A Finnish study undertaken by Kasesniemi and Rautianen demonstrates similar findings:

*The most surprising feature in the text messaging of Finnish teenagers is the extent to which it incorporates collective behaviour ... Text messages are circulated among friends, composed together and read together, and fitting expressions or entire messages are borrowed from others.* (Katz & Aakhus, 2002, p. 181)

In Weilenmann and Larsson's study, teenage boys were found to use mobile phones as a means of getting acquainted—for example, by passing the phone to a girl who then keys in her phone number so that he can ring her the following day. Weilenmann and Larsson found that the extensive sharing of the mobile phone has in a sense replaced the use of the public phone. They suggest that manufacturers need to consider teenagers' actual uses of mobile phones when designing future products and note that both Nokia and Ericsson have released mobile phones marketed in Sweden as "teenage phones". The authors note that voice recorded SMS and multi-party SMS are potential future possibilities (2002, p. 105).

## Symbolic Value

> ...the adoption of the mobile phone is not simply the action of an individual but rather it is the individuals aligning themselves with the peer culture in which they participate ...This is perhaps more true of the mobile telephone than it is with other adolescent artefacts such as clothing since the mobile is, in the first instance, an instrument with which one communicates. Beyond this the mobile telephone also communicates symbolically. The façade of the device, the type and its functions indicate something about the owner. Finally the very ownership of a mobile telephone indicates that one is socially connected. (Ling & Helmersen, 2000, p. 15)

Jon Agar, in *Constant Touch: A Global History of the Mobile Phone* (2003) describes mobile phones as "iconic markers of status," arguing that this is demonstrated in popular films such as *The Matrix* (p. 144). He notes that in 2002, global subscriptions to cellular phone services numbered more than one billion. In Iceland, Finland, Italy and the UK, more than three-fourths of the population owned a mobile phone. In the UK, 70% of adults and 81% of 15-24 year-olds owned a mobile phone.

Rich Ling's extensive research into the ownership and use of mobile telephones by Norwegian youth reveals that adoption of this communications device is widespread amongst this demographic. In a collaborative project undertaken with Per Helmerson, Ling found that "90% of Norwegian 20-year olds [have] some form of access to a mobile telephone [and] about 60% of the thirteen year-olds have some form of ownership ... [Furthermore] about 11% of nine to 10 year olds and 26% of 11-12 year olds indicate that they own a mobile telephone" (Ling & Helmersen, 2000, p. 2). The average age of mobile telephone owners within the 13-20 year-old range was approximately 14 years (Ling & Helmersen, 2000, p. 9).

As well as a communication device, the mobile phone is also an accessory—one that is assigned considerable symbolic value amongst teenagers. Its size, aesthetics, functionality and brand all contribute to this. Its display indicates to others that its owner has the economic means to support its use and may also suggest that the owner is socially in demand—"to *demonstrate* that they were a part of a social network and their status within that network" (Taylor & Harper, 2001, p. 1). Ling and Yttri found that almost all of their study's Norwegian respondents had 100-150 names listed in their mobile phones. Although many of these numbers were infrequently used, a full register was regarded by the teenagers as indicative of their social currency (2002, p. 161). The display of a mobile phone also suggests that the user is technically proficient. Alternatively, it can suggest the owner's sense of false importance. The use of the mobile phone in public places such as in a restaurant can also be seen as a form of status display for those in the immediate physical vicinity. In fact, the mobile phone also comes to form a part of the user's identity. This is particularly significant for adolescents, whose period of development is largely concerned with the establishment of individual identity (Ling, 1999, pp. 4-6).

## Mobile Phones and Parental Surveillance

*"Mum calls when I'm out drinking. Let it go and SMS her."'* (Interviewee in Carroll et al., 2000, p. 99)

*"I usually block my parent's number … but then there is a lot of hubbub when I come home."* (14 year-old interviewee in Ling & Helmersen, 2000, p. 15)

*"When you have a mobile telephone then you have a private answering machine and a private telephone."* (17 year-old interviewee in Ling & Helmerson, 2000, p. 13)

As the writer has argued previously (Quigley, 2000; Quigley & Blashki, 2003), parental concerns about the appropriation of ICT's by young people are largely based on essentialist views of childhood and technology along with boundary fears such as those surrounding the child/adult and private/public dichotomies. Moreover, the proliferation of these technologies makes it almost impossible for parents to prevent their children's access to and utilisation of them. Most importantly, the writer believes that in the post-industrial societies of the 21st century, the ability to use ICTs effectively and independently *must* be acknowledged as an essential prerequisite for the development of an active citizenry. This process is a continuum that requires that parents progressively shift their emphasis away from the protection of their children and toward their growth toward autonomy. As sociologist David Buckingham argues,

> *Children are escaping into the wider adult world—a world of danger and opportunities, in which the electronic media are playing an ever more important role. The age in which we could hope to protect children from the world is passing. We must have the courage to prepare them to deal with it, to understand it, and to become more active participants in their own right.*
> (Buckingham, 2000, p. 207)

A number of theorists have pointed out that the development of new technologies allows the young unprecedented opportunities for behaviour, along with increased power and autonomy (Barnhurst, 1998; Chapman & Chapman, 2000; Marvin, 1988). Marvin's example of the young nineteenth-century woman who used the telegraph—then a new technology—to establish relationships with members of the opposite sex without the approval of her father, is instructional in that it demonstrates that the boundary markers between parent and child have long been a site of contestation and that ICTs have served as tools in these inter-generational power struggles.

International studies all tend to show that teenagers use mobile phones as a means to escape parental control and surveillance. Ling notes that for Norwegian adolescents, the mobile phone represents "the assertion of emancipation as the device allows for social coordination unfiltered by one's parents. The possession of a mobile telephone also is the assertion of economic emancipation – even in the face of parental subsidizing of other portions of the teen's life" (Ling, 1999, p. 16). The use of text messaging also provides a means of differentiating themselves and their peers from adults (Taylor & Harper, 2001).

According to anthropologist Mizuko Ito, in Japan, where family and school exert tight control over the lives of high school students, and where the family home tends to be small and crowded, mobile phones have

> *... triggered an intergenerational power shift ... because they freed youth from 'the tyranny of the landline shared by inquisitive family members, creating a space for private communication and an agency that alters possibilities for social action' .... Texting made it possible for young people to conduct*

*conversations that can't be overheard ... [Teenagers use] this new communi-*
*cation freedom to 'construct a localized and portable place of intimacy, an*
*open channel of contact with generally three to five others'.* (Rheingold, 2002,
p. 4)

Nonetheless, at the same time as mobile phones allow young people greater privacy and freedom from parental surveillance, they also allow the coordination of children's activities by parents as well as increased security for the children themselves when away from home. A number of qualitative studies attest to this. For example, as one of the respondents in Ling and Helmerson's Norwegian study (2000)—the mother of a 17 year-old girl—explains,

*It is not to control my daughter that she should take her mobile telephone when*
*she goes out, but ... If something happens, call home and we will come*
*immediately ... Because she needs to go out and experience Oslo. She has to*
*learn about the world.* (p. 14)

Similarly, the negotiation between parents and child as to who pays the phone bills also signifies the growth towards maturation.

## Mobile Youth Communities

*In fact most contemporary communities in the developed world do not resemble*
*rural or urban villages where all know all and have frequent face-to face*
*contact. Rather, most kith and kin live farther away than a walk (or short*
*drive) so that telephone contact sustains ties as much as face-to-face get-*
*togethers.* (Wellman & Gulia in Wellman, 1999, p. 348)

*Mobile phones, unlike personal computers, are small, mobile, constantly on,*
*and potentially constantly connected.* (Brown, 2002, p. 5)

*Many of the discussions amongst the groups interviewed revealed that mobile*
*phones were valued because they were seen to foster and preserve a sense of*
*community.* (Taylor & Harper, 2001, p. 1)

Rheingold's research reveals that adolescents are creating new mobile communities and that it is 12-25 year-olds who are leading this social revolution. He cites the Swedish example of "Lunarstorm": a platform incorporating a merger of SMS text messaging and online chat which is utilised for both social and political ends—to swarm in malls or mobilize on the streets. Launched in 2000, Lunarstorm attained over one million members—representing more than 65% of Swedish 14-24 year-olds-in less than a year (Rheingold, 2002, p. 20). Mobile services have been enhanced so that anything the user can do on its Website can be done from his/her mobile phone using standard SMS messages.

The creation of mobile youth communities is also occurring in Australia. This seems to be a response to an era of change accompanied by youth pessimism and a lack of security. A 1996 study of 15-24 year-olds undertaken by the Australian Science, Technology and Engineering Council found that young people were more inclined to believe that new technologies would be used to "entrench and concentrate wealth and power" rather than to "empower people and strengthen democracy". The respondents wished for Australian society to place more emphasis on "community and family, the environment and coopera-tion" rather than "individual, material wealth and competition" (Mackay, 1997, pp. 172-173).

Like Howard Rheingold, Australian social researcher Hugh Mackay has also observed the formation of mobile youth communities. After spending the day together at school, students communicate via mobile phone on their way home, and then use their home computer to e-mail or chat. Mackay asserts that this—the "generation that beeps and hums," having grown up in an age of rapid change, rely on one another in order to cope with life's uncertainty and that in doing so, they are reforming our society.

> *Their desire to connect, and to stay connected, will reshape this society. They are the harbingers of a new sense of community, a new tribalism, that will challenge every thing from our old-fashioned respect for privacy to the way we conduct our relationships and the way we build our houses.* (Mackay, 2002, p. 7)

Carroll et al.'s Australian study also found that ICTs enable young people to overcome the fragmentation characteristic of their lives.

> *Fragmentation may arise from geographical distance as well as distinct work, educational, social and personal groups. Many of the participants talked about the different groups of people in their lives. A highly mobile population, blended families and increased numbers of overseas students have resulted in geographically dispersed contacts. Further, young people are juggling the*

*demands of other people for their time, including employers, friends, parents,*
*teachers, sports teams and coaches. ICTs such as email, mobile phones and*
*chat help young people to develop and maintain virtual communities of family,*
*friends and other young people with similar interests.* (Carroll et al., 2002, p. 99)

It must, of course, also be acknowledged that not everyone utilises new technologies in benevolent ways. Bettina Arndt notes that there is a significant black market operating amongst Australian teenagers who are involved in stealing or trading mobile phones. This in itself signifies the highly desirable nature of this device (2003, p. 6). During their data collection, Kasesniemi and Rautianen found that some of their respondents had admitted to using their mobile phones to supply alcohol to the under-aged, for stealing and shoplifting (2002, p. 174). Nonetheless, any increase in an adolescent's freedom is bound to result in some misdemeanours, the extent of which may indicate their level of maturation as well as the nature of their moral upbringing. Overall, however, the research makes abundantly clear that young people's use of ICTs is primarily concerned with fostering and strengthening communication and community.

## Conclusion

Long regarded as in the forefront of new technology appropriation and use, young people in developed countries are demonstrating how technologies are adapted to social needs. The rapid and ubiquitous appropriation of mobile phones by young people in recent years is based on a number of factors including security, self-identity, privacy and freedom—and, most importantly—because mobile phones allow the reinforcement or extension of their existing face-to-face social networks. Rather than being regarded, on the one hand, as a trivial device, or, alternatively, as a tool that is used primarily to enable premature escape from parental surveillance, a number of international studies show that the mobile phone is generally utilised by the young in socially responsible ways. They are reclaiming the "lost paradise" of community in the twenty-first century by means of their creative adaptation of new information and communication technologies. At a time of rapid and significant social change, they utilise these technologies to form and to maintain mobile social networks or communities that combine virtual and face-to-face communication between established relationships and new ones. In doing so, they simultaneously maintain a sense of self-identity and place, freedom and belonging, thereby demonstrating their development towards citizenship as well as offering new possibilities for community formation and renewal within electronic societies.

# References

Agar, J. (2003). *Constant touch: A global history of the mobile phone.* Cambridge, UK: Icon Books.

Anderson, B. (1991). *Imagined communities: Reflections on the origin and spread of nationalism.* London: Verso.

Arndt, B. (2003, August). Cell mates. *Age*, A3, 6.

Bakardjieva, M. (2003). Virtual togetherness: An everyday-life perspective. *Media, Culture & Society, 25,* 291-313.

Barnhurst, K.G. (1998). Politics in the fine meshes: Young citizens, power and media. *Media, Culture and Society, 20* (2), 201-18.

Boethius, U. (1995). Youth, the media and moral panics. In J. Fornas & G. Bolin (Eds.), *Youth culture in late modernity* (pp. 39-57). London: Sage Publications.

Brown, B., Green, N., & Harper, R. (2002). Wireless world: Social and interactional aspects of the mobile age. In Diaper (Ed.), *Computer supported cooperative work.* London: Springer-Verlag.

Buckingham, D. (2000). *After the death of childhood: Growing up in the age of electronic media* (1st ed.). Cambridge, MA: Polity.

Carroll, J. et al. (2001). Identity, power and fragmentation in Cyberspace: Technology appropriation by young people. *ACIS 2001: Proceedings of the Twelfth Australasian Conference on Information Systems* (pp. 98-99). Southern Cross University.

Chapman, N., & Chapman, J. (2000). *Digital multimedia.* Chichester: John Wiley and Sons.

Heller, A. (1984). *Everyday life.* London: Routledge and Kegan Paul.

http://parlsec.treasurer.gov.au/parlsec/content/pressreleases/2002/033.asp?pf+1

Kasesniemi, E., & Rautiainen. (2002). Mobile culture of children and teenagers in Finland. In J.E. Katz & M. Aakhus (Eds.), *Perpetual contact: Mobile communication, private talk, public performance.* Cambridge: Cambridge University Press.

Kendall, L. (2002, January). I just texted to say I love you. *Advertiser,* 11.

Ling, R. (1999, July). "We release them little by little": Maturation and gender identity as seen in the use of mobile telephony. International Symposium on Technology and Society (ISTAS'99). *Women and Technology: Historical and Professional Perspectives.* Rutgers University, New Brunswick, New Jersey.

Ling, R., & Helmersen, P. (2000, June). *"It must be necessary, it has to cover a need": The adoption of mobile telephony among pre-adolescents and adolescents.* Presented at the Conference on the Social Consequences of Mobile Telephony. Oslo, Norway.

Ling, R., & Yttri, B. (2002). Hyper-coordination via mobile phones in Norway. In J. Katz & M. Aakhus (Eds.), *Perpetual contact: Mobile communication, private talk, public performance.* Cambridge, UK: Cambridge University Press.

Mackay, H. (1997). *Generations: Baby boomers, their parents and their children.* Sydney: Macmillan.

Mackay, H. (2002). Our new way of being a community. *Age: Melbourne,* 7.

Marvin, C. (1988). *When old technologies were new: Thinking about electric communication in the late nineteenth century.* New York: Oxford UP.

Meyrowitz, J. (1997). The separation of social space from physical place. In T. O'Sullivan et al. (Eds.), *The media studies reader* (pp. 42-52). London: Edward Arnold Ltd.

Meyrowitz, J. (1999). No sense of place: The impact of electronic media on social behaviour. In H. Mackay & T. O'Sullivan (Eds.), *The media reader: Continuity and transformation* (pp. 99-120). London: Sage.

Postman, N. (1987). *Amusing ourselves to death: Public discourse in the age of show business.* London: Methuen.

Postman, N. (1994). *The disappearance of childhood.* New York: Vintage Books.

Quigley, M. (2000). The politics of animation: South Park. *Metro*. St Kilda: Australian Teachers of Media, *124/125*, 48-54.

Quigley, M., & Blashki, K. (2003). Beyond the sacred garden: Children and the Internet. *Information Technology in Childhood Education Annual*. Association for Advancement of Computing in Education, 309-316. This article also appears in *Educational Technology Review*, [Online], *11*(1). Available: http://www.aace.org/pubs/etr

Rheingold, H. (2000). *The virtual community: Homesteading on the electronic frontier* (rev. ed.). Cambridge, MA: MIT Press.

Rheingold, H. (2002). *Smart mobs: The next social revolution*. Cambridge, MA: Perseus.

Silverstone, R. (1999). *Why study the media?* London: Sage.

Strathern, M. (1992). Foreword: The mirror of technology. In R. Silverstone & E. Hirsch (Eds.). *Consuming technologies: Media and information in domestic spaces* (p. x). London: Routledge.

Taylor, A.S. & Harper, R. (2001). Talking "activity": Young people & mobile phones. *CH1 2001 Workshop: Mobile communications: Understanding users, adoption & design*.

Weilenmann, A., & Larsson, C. (2002). Local use and sharing of mobile phones. In B. Brown, N. Green & R. Harper (Eds.), *Wireless world: Social and interactional aspects of the mobile age*. London: Springer.

Wellman, B. (Ed.). (1999). *Networks in the global village*. Boulder, CO: Westview Press.

Williams, R. (Ed.). (1981). *Communications technologies and social institutions. Contact: Human communication and history*. London: Thames and Hudson.

Williams, R. (1985). *Keywords: A vocabulary of culture and society*. New York: Oxford University Press.

# Part II

---

# Information Security

Chapter IX

# Insights from Y2K and 9/11 for Enhancing IT Security[1]

Laura Lally
Hofstra University, USA

## Abstract

*In the post-9/11 environment, there has been an increasing awareness of the need for information security. This chapter presents an analysis of the Y2K problem and 9/11 disaster from the perspective of Lally's extension of Perrow's Normal Accident Theory and the Theory of High Reliability Organizations. Insights into: 1) how characteristics of current IT infrastructures and organizational cultures make disasters more likely, 2) how organizations can respond to potential threats and mitigate the damage of those that do materialize, and 3) how IT can be used to identify future threats and mitigate their impact in the future, emerge from the analysis.*

# Introduction

In the post-9/11 environment, information technology managers have become more aware of the importance of security. Throughout the 1990s, IT security faced a wide range of new challenges. Yourdon (2002) places these challenges in three categories:

1.  More organizations are dependent on the Internet for day-to-day operations.

2.  An increasing number of computer systems, networks and databases make up a global IT infrastructure. Individuals, organizations and nations are "increasingly 'wired,' increasingly automated, and increasingly dependent on highly reliable computer systems" (Yourdon, 2002, p. 96).

3.  IT managers faced more sophisticated and malevolent forms of attacks on these systems. Unlike the Y2K problem, which was the result of an innocent bad judgement, "the disruptive shocks to our organizations are no longer accidental, benign, or acts of nature; now they are deliberate and malevolent" (Yourdon, 2002, p. 205).

This chapter will present an analysis of the sources, propagation and potential impacts of IT related threats. The Y2K problem and the information technology implications of 9/11 will be used to illustrate the analysis. The analysis will focus on both: 1) how the current IT infrastructure allows for the propagation of IT based threats, and 2) ways in which available IT tools can help identify potential threats and mitigate their impact.

# Extending Perrow's Normal Accident Theory and the Theory of High Reliability Organizations

This analysis will draw on Lally's (2002) extension of Perrow's Normal Accident Theory (1984, 1999), as well as the Theory of High Reliability Organizations. Perrow developed his theory studying complex systems such as nuclear power plants. He distinguished characteristics of systems that would permit single failures, called "incidents" such as an operator error, to propagate into major accidents such as meltdowns. Systems that had these characteristics were likely to be subject to accidents in the normal course of their operation. Perrow concluded that accident prone systems are more:

1)   *Complex*—with only the "Tip of the Iceberg" visible, leading to the problem of *"unknowability,"*

2)   *Tightly coupled*—with no slack time to allow incidents to be intercepted, and

3)   *Poorly controlled*—with less opportunity for human intervention before problems spread.

Lally (1996) argued that Normal Accident Theory is a sound theoretical perspective for understanding the risks of information technology because IT is:

1)   *Complex*—The hardware that makes up IT infrastructures of most organizations is complex, containing a wide range of technologies. Software often contains thousands of lines of code written by dozens of programmers. Incidents such as bugs can, therefore, propagate in unexpected ways;

2)   *Tightly coupled*—Both hardware and software are designed to increase the speed and efficiency of operations. Incidents such as operator errors can quickly have real-world impacts; and

3)   *Poorly controlled*—Security features are often not built into systems. Testing of software is often inadequate in the rush to meet release deadlines.

## High Reliability Theory – A Complementary Perspective

Researchers in the Theory of High Reliability Organizations have examined organizations in which complex, tightly coupled, technologically based systems appeared to be coping successfully with the potential for disaster. Their studies of the Federal Aviation Administration's air traffic control system, the Pacific

Gas and Electric's electric power system, including the Diablo Canyon nuclear power plant, and the peacetime flight operations of three United States Navy aircraft carriers indicate that organizations can achieve nearly error free operation (La Porte & Consolini, 1991; Sagan, 1993).

High reliability organization theorists identify four critical causal factors for achieving reliability: 1) Political elites and organizational leaders put safety and reliability first as a goal; 2) High levels of redundancy in personnel and technical safety measures; 3) The development of a "high reliability culture" in decentralized and continually practiced operations; and 4) Sophisticated forms of trial and error organizational learning.

The two theories have been contrasted as "pessimistic" — Perrow's contention that disaster is inevitable in badly designed systems, versus "optimistic" — La Porte's pragmatic approach to achieving greater reliability. The theories, however, are in agreement as to which characteristics of systems make them more or less accident-prone.

Lally applied these theories to various aspects of information technology, including reengineering (Lally, 1996, 1997), the Y2K problem (Lally, 1999), and privacy in the hiring processes (Lally, 2000; Lally & Garbushian, 2001). Lally concluded (Lally, 2002) that the rapid pace of *change* in information technology is a further exacerbating factor increasing the likelihood of disasters:

1) *Changes in Hardware*—According to Moore's Law, hardware doubles in power every 18 months. As a result, hardware continues to evolve rapidly. Furthermore, entirely new kinds of hardware appear and must be integrated into existing systems;

2) *Changes in Software*—New software releases fuel revenue streams in the software industry, resulting in mandatory "upgrades" every two years. The changes create an additional learning burden on users. Programmers are again under time pressure that can result in poor testing and debugging (Austin, 2001; Halfhill, 1998; Westland, 2000).

In addition to these first order effects, Lally (1997) also argues that changes in IT create second order effects by enabling changes in organizational processes. These processes can also become more complex, tightly coupled, and poorly controlled, further increasing the problem of serious accidents. As a result, IT managers and users are faced with complex, tightly coupled, poorly controlled systems that undergo radical changes on a regular basis, making these systems more prone to "Normal Accidents".

# Y2K from Design Flaw to Potential Disaster

Y2K was caused by short-sightedness on the part of software designers, who were constrained by a lack of space on Hollerith punch cards and in the computers' main memory in the 1950s. As a result, the penny-wise, pound-foolish decision was made to use only two significant digits to represent the year in the date field. The use of two character year fields caused inappropriate representations of the year 2000 in a wide range of applications, causing transactions to fail, computer controlled machines to malfunction, and computations involving algorithms with date functions to provide wrong answers. By mid-1999, the Information Technology Association of America reported that 44% of the U.S. companies they polled had already experienced Y2K related failures (Poulson, 1999). Experts differed widely in their estimate of the likelihood of serious system failures and the consequences of these failures, ranging from predictions of minor disruptions to worldwide recession and civil unrest (Anson, 1999).

Y2K was the first computer related risk that received the commitment of major resources from corporate CEOs and heads of government. This chapter will argue that Y2K was merely a symptom of a more generic problem in computer information systems—that, by the nature of their current design, they will be prone to more problems which may have more serious long-term consequences than Y2K did.

## Characterizing Y2K as an Incident in a Single Organization

*The Millenium Horse is a virus hacker's best dream (and our worst nightmare). It is absolutely everywhere. It is "polymorphic," having a nearly infinite variety of structures and behaviors, and in a nanosecond it can destroy your data, your business and perhaps your life.* (Lefkon, 1997, p. 70)

This chapter will argue that Y2K illustrated many of the characteristics of a problem in a Normal Accident environment.

1. Incidents such as a mismatched date representation were difficult to predict in advance and companies were slow to address the problem.

2. The complex, tightly coupled nature of computer systems made the propagation of these incidents difficult to assess.

3. The control mechanisms available for dealing with the incidents that typically arise in a computerized environment did not work.

Furthermore, Y2K was considered likely to produce a large number of incidents over a short period of time, making the potential for serious disruptions severe.

Perrow (1999, p. 392) argued, "Y2K has the potential for making a linear, loosely coupled system more complex and tightly coupled than anyone had reason to anticipate". Perrow emphasized that Y2K made him more keenly aware of the problem of "unknowability":

> One of the key themes of the theory, but not one formalized as much as it could have been, was the notion of incomprehensibility—What has happened? How could such a thing have happened? And What will happen next? This indicated that observers did not know what was going on and did not know what would happen next. The system was in an unknowable state. (Perrow, 1999, p. 293)

## Why the Delay in Getting Started

Y2K as a general systems problem was realized as early as 1984 when a *Computerworld* article recognized the problem in COBOL code. Why then did CEOs take so long to marshal the resources to address it?

Pauchant and Mitroff (1992) indicate that many CEOs are typically in denial about the potential for crises in their organization. Lefkon (1998) went so far as to suggest Elizabeth Kubler Ross's four-step cognitive model for resigning oneself to death for CEOs who have delayed beginning Y2K efforts. On a less gloomy note, Nemeth, Creveling, Hearn and Lambros (1998) suggested that the emotional factors associated with a proactive approach to Y2K problems were:

1.  A high level of knowledge.
2.  A low level of denial.
3.  Manageable levels of discouragement.
4.  Manageable, but slightly higher, levels of apprehension.

Nemeth et al. (1998) indicated that managers who linger too long in denial are likely to fall directly into a state of disabling discouragement.

High-level managers rarely have an understanding of the technical risks of computer systems. For CEOs to recognize a risk, the risk must be "framed" in terms of the types of risks they are familiar with. Therefore, it took the first lawsuit brought by a retailer against the vendor of a credit card authorization system that rejected a credit card with a "00" expiration date to awaken the banking community in 1995 to the potential of disastrous consequences. Other industry segments had not received such legal wake-up calls because the date problem and attending lawsuits had not yet arisen in their normal course of business.

Insights from High Reliability Organization theory can help explain the differences in response to Y2K across industries. In financial services, high-level regulatory authorities such as the Federal Reserve and the Federal Financial Institution Council took a proactive approach to ensuring compliance. The Federal Reserves program included:

1)  "shared testing" weekends where member institutions could test their mission critical applications;
2)  an automated inventory of the vendor products in use; and
3)  independent tests of software.

As a result, all critical financial services systems in the U.S. were compliant by the deadline. The presence of an independent regulatory agency with the power to punish non-compliant institutions appeared to be an effective means of creating a high reliability culture regarding Y2K.

In the area of electrical power distribution, an organization of power providers, the North American Reliability Council (NERC), worked on coordinating the electric power industry's response to Y2K. NERC, however, lacked the elite status and clout of the Federal Reserve. Perhaps this is why only 44% of electric power entities reported that testing would be completed as of June 1999. As of November 1998, only 72% had written Y2K plans. In natural gas distribution, there was no single, coordinating body, like NERC, leading the way. Self reported data were incomplete and vague. In Europe, power companies were even later getting started. In March 1998 a simulated Y2K test in Hanover, Germany caused a number of systemic failures that took seven months to completely isolate and correct (Andrews, 1999). This event resulted in the German government initiating a series of Y2K programs.

In telecommunications, the overwhelming complexity of the Public Switched Telephone Network (PSTN) made 100% testing impossible. Compounding the problem were the thousands of privately owned networks attached to the PSTN and the lack of an independent controlling authority.

## Why the Difficulty in Fixing the Problem

The tightly coupled, complex nature of software made isolating and containing Y2K errors difficult. Software, the instructions that run computers, generally consists of thousands of lines of code for a major application. Programs generally consist of many modules that interact with each other in a non-linear fashion. The code is generally developed by a number of programmers, many of whom never see the entire program:

> *Most programming is done through what are called application programming interfaces or APIs....The code on the other side of the interface is usually sealed in a proprietary black box. And below that black box is another, and below that another—a receding tower of black boxes, each with its own errors. You can't envision the whole tower, you can't open the boxes, and what information you've been given about any individual box could be wrong.* (Ullman, 1999, p. 3)

Additionally, programmers are often likely to be unaware of real-world impacts of the program. One programmer who worked for the Federal Reserve System only became aware of what the Federal Reserve actually did when he joined their Y2K team: "I read an article about how the Federal Reserve would crash everything if it went bad…I discovered we were kind of important" (Ullman, 1999, p. 4).

Conversely, the business managers who rely on the software to run their business are generally blocked from reading the code in a purchased package, and are unable to read the code developed in-house. Interactions in computer systems are generally non-linear and invisible, so the manner in which errors propagate is not visible without a thorough examination of the code and system documentation. Testing routines and software specially designed for the purpose must be used to trace error propagation.

Software development methodologies attempt to minimize the likelihood of programming errors (called "bugs") by testing, but developers admit the near impossibility of eliminating all of them. Methodologies have evolved to ensure high degrees of reliability in complex software considered "mission critical" — a failure could prevent the system from functioning. These methodologies, however, greatly increase development time and costs (Leveson, 1995). Business software, developed under greater economic constraints, rarely places such a high dividend on reliability, engaging instead in "aggressive scheduling". "Release schedules are normally driven by market demand, not the actual time it would take to build a reasonably robust system. The parts of the development process most often foreshortened are two crucial ones: design documentation and testing" (Ullman, 1999, p. 5). In fact, software in business is often accompanied by disclaimers relieving developers of any legal liability resulting from the harmful impacts of their errors.

> *Imagine your reaction if you found that disclaimer on your car or kitchen appliance. What would happen to our society if everybody who wished to use a telephone, television set, car, detergent of plastic toy were first obliged to learn at least a little about how it was made and how it works internally, and then to test it for hazards and other surprises? Why are software manufacturers allowed a sweeping disclaimer that no other manufacturer would dare to make?* (Peterson, 1995, p. xvii)

Critics of business software development call for increased degrees of accountability and legal liability and for instituting licensing and performance standards (Nissenbaum, 1994).

The impact of bugs is exacerbated by the tightly coupled nature of software modules. Errors propagate through the code in fractions of a second and can have real-world impacts in seconds. Computer processing chips grow exponentially more complex as well. Hardware platforms are increasingly powerful, fast, and integrated with one another through high-speed telecommunication links, creating even tighter coupling and less opportunity to control the propagation of errors through human intervention.

Berghel (1998) characterized Y2K as a classic example of the "Riddle of Induction" first posed by 18th Century philosopher David Hume. Hume criticized all inductive reasoning on the grounds that unless one understands the true relationship between cause and effect, one can only have an instinctive understanding of the system that produces the effect and never be sure that new situations will not cause the system to behave differently. Berghel extrapolated this to the testing of computer systems, stating that if one only tests inputs and outputs without examining the code, one cannot be sure that a new situation, such as Y2K, will not cause a malfunction. Computer systems are too complex for users to understand and, in the case of embedded chips, are often too ubiquitous for users to even know they are there.

The changes undergone by computer systems since the first programs were written also exacerbated the problem. The original programmers often did not use meaningful, self documenting, variable names. Variables that could have been called "Loan-Due-Date" were instead named "Six-Character Field". Programmers also neglected to document programs or removed documentation after completing the program to ensure further employment.

> *One is Hal Varian, now dean of the School of Information and Management Systems at the University of California at Berkeley. In 1966 Professor Varian worked for a big Boston defense contractor. "At the end of summer," he reminisces, "I was going back to school, and my boss was also leaving the company. On my last day, he told me to go through the assembly-language programs I had spent the summer working on and take out all the comment cards. 'But then no one will be able to understand the program,' I protested. 'Exactly,' he said. 'They'll have to hire us both back as consultants if they want to make any changes'.* (Cairncross, 1998, p. 1)

In the case of packaged software, the code is proprietary, and therefore cannot be read by its users. Vendors of large reputable systems, such as Microsoft and IBM, diagnosed their own software and provided patches or new versions that addressed the problem. These vendors also used Websites to keep users informed of the compliance status of their products. Less reputable firms, however, declared bankruptcy rather than deal with the Y2K problem, leaving many users with systems they could not test.

# The Potential for Global Disasters:
# How Y2K Problems Could Have Propagated Beyond
# Organizational Boundaries

Computer systems are increasingly integrated and span organizational boundaries. Telecommunication based order systems from suppliers and delivery systems to customers are becoming more common. The Internet, a vastly complex, highly interconnected platform has become a widely used business tool for integration both within and between businesses. All these provided a means by which Y2K failures in one organization could have propagated to other organizations and to the economy as a whole.

Y2K practitioners referred to this as the "ripple effect" (Kirsner, 1997), based on their observations that few computers operate in stand-alone environments. Golter and Hawry (1998) characterized six "circles of risk" that IS managers must address:

1)    core information systems;
2)    networks, PCs and workstations;
3)    exchanges with third party vendors;
4)    plant and equipment;
5)    business partners; and
6)    the macro economy.

In a tightly coupled environment, failures in any of these areas could radiate outward before they could be contained.

On the corporate level, practitioners Beach and Oleson (1998) developed a "Pain index" of how serious the ripple effect would be, depending on how many connected applications an organization has. With a stand-alone system, the risk is zero; with 1500 applications the risk of a catastrophic failure rises to 15.2% and the risk of annoying failures reaches 100%.

Managers were warned by Y2K practitioners that if their entire supply chain was not Y2K compliant then their organizations could be subject to system-wide failures propagating into their organization from Y2K incidents generated externally (Nolan, 1998). Organizations that had integrated their supply chain to eliminate paperwork were ironically forced to create a huge paper flow of letters insisting that their trading partners be Y2K compliant.

On a global level, the lack of compliance by many countries led airlines to consider practicing a form of fault isolation, establishing "no-fly zones" over non-compliant countries (Anson, 1999). Concern was also expressed about potential cross-border damage between countries that were Y2K compliant and those that were not. As a result of many countries' non-compliance, Ed Yardeni, Chief Economist at Deutsche Morgan Grenfell predicted that there was a 60% chance that Y2K would lead to a global recession (Golter & Hawry, 1998).

## Why No Global Disasters

This chapter will argue that Y2K did not result in global disaster because, although there were many local failures, they did not propagate globally, and the "ripple effect" did not occur. This was primarily a result of Y2K being an innocent error. Y2K was not a deliberate "incident;" programmers who created the error were not trying to cause damage. Programmers did not use their knowledge of computer systems to maximize the propagation of the problem to cause damage. IT based systems, however, are also threatened by a number of deliberate attacks.

Computer based attacks such as viruses and worms, however, are malicious and deliberately designed for maximum propagation, unlike Y2K, which was an inadvertent error. Perpetrators of these attacks often have inside knowledge of security flaws that will allow malicious incidents to propagate further. Other types of threats exist as well.

## 9/11/01 – What has happened?
## How could such a thing happen?
## What will happen next?

On September 11, 2001, a surprise terrorist attack left the world wondering, "What has happened? How could such a thing happen? What will happen next?" The damage caused by the initial impact of the planes quickly spread, destroying the World Trade Center and causing massive destruction and loss of life in lower Manhattan.

The Y2K problem was an innocent error, recognized ahead of time, and prevented from causing catastrophic failures. 9/11 was a deliberate, well organized, surprise attack that caused catastrophic damage before the military, the police, or the thousands of individuals who lost their lives could do anything to circumvent it. The prediction of the Y2K bug causing a worldwide recession did not come true. 9/11, however, will have serious global economic ramifications for years to come.

"Responding to terrorism will be a more complex task,"…as John Koskinen, former head of the government's Y2K effort, remarked recently, "Unlike the Y2K Phenomenon, today's terrorist threat to IT is undefined, the response is difficult, and there is no known time frame" (Yourdon, 2002, p. 29).

A number of parallels, however, do emerge between the two events that can provide insight for preventing and/or mitigating the impacts of future terrorist attacks. Both emphasized the importance of planning for catastrophic failures. Some organizations indicated that their Y2K planning helped them mitigate the damage caused by 9/11 (Merian, 2001). Pressure is on from the business community to re-create the U.S. government effort in combating the Y2K problem as a means of combating terrorism (Thibodeau, 2001). This chapter will argue that Y2K, therefore, provides a useful starting point in analyzing the 9/11 disaster.

## Recognizing an Incident and Understanding its Potential Impact

From a Normal Accident Theory perspective, a number of critical issues emerge regarding 9/11. First, what was the "incident" that needed to be recognized? Was it: 1) The first plane hitting the North Tower—at which point a serious problem became "knowable"? 2) The second plane hitting the South Tower—at which point a terrorist attack could be identified as occurring? At this point there was no need to convince anyone that a serious problem existed. Here we can clearly distinguish between the "intellectual" threat of Y2K, which required large numbers of technical experts to convince the public of its seriousness and the "visceral" threat experienced by anyone viewing the 9/11 disaster. This chapter will argue that although the plane crashes propagated into even greater destruction, the first plane crash was already an "accident" leading to massive destruction and loss of life. Incidents preceding this event, if intercepted, could have prevented its occurrence. Examples of such incidents include:

1) terrorists boarding the planes,

2) discovering the existence of the 9/11 plot,

3) hearing a young man say he wishes to learn how to steer an airliner, but not how to take off and land.

## Why weren't the incidents contained?

Similarly to Y2K, the potential risk of a 9/11 type attack was recognized but not acted upon. The Report of the Joint Inquiry into the Terrorist Attacks of September 11, 2003—by the house Permanent Select Committee on Intelligence and the Senate Select Committee on Intelligence (2002) indicated a number of incidents which had been recognized and recorded at the local level but not acted upon in time and in a manner to prevent the attacks:

1. The Intelligence Community had amassed a great deal of valuable intelligence regarding Osama Bin Laden and his terrorist activities…that was clearly relevant to the September 11 attacks, particularly when considered for its collective significance.

2. Beginning in 1998 and continuing into the summer of 2001, the Intelligence Community received a modest, but relatively steady, stream of intelligence reporting that indicated the possibility of terrorist attacks within the United States.

3. From at least 1994, and continuing into the summer of 2001, the Intelligence Community received information indicating that terrorists were contemplating, among other means of attack, the use of aircraft as weapons. This information did not stimulate any specific Intelligence Community assessment of, or collective U.S. government reaction to, this form of threat.

4. The Intelligence Community acquired additional, and highly significant information regarding Khalid al-Mihdhar and Nawaf al-Hazmi in early 2000. Critical parts of the information concerning al-Mihdhar and al-Hazmi lay dormant within the Intelligence Community for as long as

eighteen months, at the very time when plans for the September 11 attacks were proceeding. The CIA missed repeated opportunities to act based on information in its possession that these two Bin Ladin associates were traveling to the United States and to add their names to the watch lists.

5.    ...the same two future hijackers, Khalid al-Mihdhar and Nawaf al-Hazmi, had numerous contacts with a long-time FBI counterterrorism informant.

6.    An FBI Phoenix filed office agent sent an "Electronic Communication"...the agent expressed his concerns, based on his first-hand knowledge, that there was a co-ordinated effort underway by Bin Laden to send students to the United States for civil aviation-related training. He noted that there was an "inordinate number of individuals of investigative interest" in this type of training in Arizona and expressed his suspicion that this was an effort to establish a cadre of individuals in civil aviation who would conduct future terrorist attacks....However, the FBI headquarters personnel did not take the action requested by the Phoenix agent prior to September 11, 2001.

The report faults the Intelligence Community for failing to use information technology tools that were available at the time to combine the information received about the incidents to create a comprehensive characterization of the potential attack. The report also faults the agencies' work processes, which inhibit, rather than enhance, the sharing of information.

> *The Intelligence Community failed to capitalize on both the individual and collective significance of available information that appears relevant to the events of September 11. As a result, the Community missed opportunities to disrupt the September 11th plot by denying entry to or detaining would be hijackers, to at least try to unravel the plot through surveillance and other investigative work within the United States, and, finally, to generate a heightened state of alert and thus harden the homeland against attack.*

Among the causes of the lack of response was:

1.    While technology remains one of this nation's greatest advantages, it has not been fully and most effectively applied in support of U.S. counterterrorism efforts. Persistent problems in this area included a lack of collaboration between Intelligence Community agencies, a reluctance

to develop and implement new technical capabilities aggressively, the FBI's reliance on outdated and insufficient technical systems, and the absence of a central counterterrorism database.

2.    ...The Intelligence Community's ability to produce timely and significant signals intelligence on counterterrorism was limited by NSA's failure to address modern communications technology aggressively, continuing conflict between Intelligence Community agencies, NSA's cautious approach to any collection of intelligence relating to activities in the United States, and insufficient collaboration between NSA and the FBI regarding the potential for terrorist attacks in the United States.

High Reliability Theory addresses these problems. It recommends the creation of a decentralized High Reliability culture in which individuals are encouraged to report information they consider threatening. "..we need to make it easier for front line observers to communicate their warnings quickly and effectively, without worrying about being criticized as alarmists" (Yourdon, 2002, p. 199).

Isenberg's theory of social coherence (Isenberg, 1999), in which individuals and organizations co-operate in sharing information, also supports this approach. Isenberg argues that if individuals can work together in response to an incident they can respond more effectively to contain it. Work processes, supported by information technology can enhance information sharing as long as the culture has changed to permit it.

More sophisticated IT based methods are also becoming available (Verton, 2002). Customer relationship management software, such as that used by Amazon books to detect patterns in buyer behavior, can also be used to detect patterns of suspicious behavior. If initial fears are confirmed, collaborative projects can help make early warnings more widely available. Sophisticated database technology can allow for the sharing of information between agencies. Facial recognition software can help identify "watch list" individuals at airports and other critical facilities.

## Modeling the Unthinkable: Mitigating the Impact of Terror Attacks

On 9/11, many lives were lost after the initial impact of the two planes because bad judgments based on incomplete information were made by individuals working in the towers, as well as by firefighters and police. This was particularly true in the North Tower. Individuals remained in the upper portion of the North Tower, hoping to be rescued despite the fact that they were unreachable and that one stairway was still passable. Firefighters continued climbing up into the lower portion of the North Tower despite the fact that the South Tower had collapsed and they were in imminent danger. Incompatible communication devices prevented critical information from reaching the firefighters (Dwyer, 2002). However, wireless communication devices, e-mails and other Internet communication did increase social coherence during the disaster. Victims said goodbye to loved ones, and the passengers on Flight 93 were able to mitigate the impact of the disaster they had become a part of. Regular business communication took place over employee cell phones and personal Internet accounts (Disabatino, 2002; Kontzer, 2002). Since 9/11, communication standards have improved between emergency service workers to increase social coherence.

Simulation models of the building, such as those designed afterward (see Nova's "Why the Towers Fell") could be used to minimize the problem of "unknowability" that resulted in so many deaths. Office workers in all large complex buildings could use these models to develop optimal evacuation plans during an emergency. Firefighters could train by simulating rescue missions in all large complex buildings in their area. Finally, in terms of social coherence, good communication between structural engineers, who are best able to determine the condition of the building, and the workers and firefighters inside could also save hundreds of lives. Since 9/11 many corporations have increased emergency training for their employees.

# Conclusion:
## The Need for Systemic Solutions

*The millenium bug is not unique; it's just the flaw we see now, the most convincing evidence we have of the human fallibility that lies inside every system. It's hard to overstate how common bugs are. Every week, the computer trade paper* Infoworld *prints a little box called, "The Bug Report," showing problems in commonly used software, some of it very serious.* (Ullman, 1999)

Y2K and 9/11 were wake up calls to business managers about the vulnerability of their systems. Terms such as "mission critical," once primarily used by risk analysts, the military and NASA are now being used in the day-to-day environment of business. This new awareness should lead to system designers and business managers taking concrete steps to prevent future failures.

Both Y2K and 9/11 spawned special legislation to address what were considered to be out-of-the-ordinary problems. This chapter argues, however, that systemic rather than legalistic solutions are needed because many out-of-the-ordinary problems are possible.

- *Develop more rigorous standards for software development.* These should include: 1) The common representation of other fields, beside date fields, that allow for the sharing of data between systems; 2) more rigorous standards for documentation and testing; and 3) a fuller awareness on the part of programmers as to the overall functioning and real-world impacts of their programs. Licensing and certification of programmers would help ensure their qualifications. Software vendors should not expect legislative shields against mission critical errors in their software. Higher levels of accountability are necessary. In the post-9/11 environment, greater data sharing between government agencies is being used a tool to combat terrorism. These complex, tightly coupled databases will require careful control against incidents such as incorrectly entered data.

- *An examination of organizational processes enabled by information technology.* Both system designers and business managers need to understand and document all complex, tightly coupled systems with the aim of making them safer and more resilient. Redundancy should be built into critical systems. In the event of an incident, systems should be able to be decoupled. Automated and human control must be designed to anticipate a wide range of potential incidents and to contain their propagation. In the case of government agencies in the post-9/11 environment, new IT supported processes must be designed to enhance data sharing and collaboration.

- *Organizations must develop a High Reliability culture.* High-level managers must recognize and express their support for making systems safe and reliable. CEOs and their organizations need to leverage their Y2K costs by viewing the experience of achieving compliance as a form of "sophisticated trial and error learning" advocated by the Theory of High

Reliability Organizations. The comprehensive software review required for Y2K compliance, the identification of mission critical systems, and the evolution of testing tools and methodologies have all helped remove the shroud of ubiquity that has cloaked the information infrastructures of most organizations. Critical industries such as electric power, gas and telecommunications should establish permanent independent governing entities to ensure a high awareness of, and accountability for, computer related as well as terrorist risks. Above all, the increased level of knowledge of an organization's vulnerabilities and exposures to risks should inform and direct the development of more reliable systems in the future.

# Endnote

[1]   This research was sponsored by a Summer Research Grant from the Frank G. Zarb School of Business at Hofstra University.

# References

Andrews, E. (1999). Europe Rides Bumpy Computer Road to Year 2000. *New York Times,* July 23, 1999, A3.

Anson, R.S. (1999). 12.31.99. *Vanity Fair,* January, 80-84.

Austin, R. (2001). The Effects of Time Pressure on Quality in Software Development. *Information Systems Research,* June, pp. 195-207.

Beach, G. & Oleson, T. (1998).  The Beach/Oleson Pain Index. *Computerworld,  http://www.cio.com/marketing/releases/ y2kchart.html.*

Berghel, H. (1998). The Year 2000 Problem and the New Riddle of Induction. *Communications of the ACM, 41*(3), 13-17.

Cairncross, F. (1998). The Art of Bug Hunting. from Survey: The Millenium Bug. The Economist, September 19, 1998, *http://www.economist.com/ editorial/freeforall/19-9-98survey/bug3.html..*

Disabatino, J. (2001). Internet Messaging Keeps Businesses, Employees, in Touch. *Computerworld,* September 17.

Dwyer, J. (2002) Radio Problem Could Last Years. *New York Times,* September, 18.

Golter, J. & Hawry, P. (1998). *Circles of Risk. http://year2000.com/ archive/circlesrisk.html.*

Halfhill, T. (1998) Crash-Proof Computing. *BYTE www.byte.com/art/9804/ sec5/art1.html.*

Isenberg, D. (1999). SMART Letter $16. *www.isen.com,* February 1.

Kirsner, S. (1998). The Ripple Effect. *http://www.cio.archive/ y2k_ripple_content.html.*

Kontzer, T. (2001). With Phone Lines Bottlenecked, Internet Messaging Became Lifeline. *Information Week,* September, 12.

Lally. L. (1996). Enumerating the Risks of Reengineered Processes. *Proceedings of 1996 ACM Computer Science Conference,* 18-23.

Lally, L. (1997). Are Reengineered Organizations Disaster Prone? *Proceedings of the National Decision Sciences Conference,* pp. 178-182.

Lally, L. (1999). The Y2K Problem: Normal Accident Theory and High Reliability Organization Theory Perspectives. *Proceedings of the 1999 National Decision Sciences Conference,* pp. 234-237.

Lally, L. (2000). Pre-Employment Screening for Reengineered Jobs: Issues in Information Access and Information Privacy. *Proceedings of the National Decision Science Conference,* 2000, pp. 371-373.

Lally, L. (2002). Complexity, Coupling, Control and Change: An IT Based Extension to Normal Accident Theory. *Proceedings of the International Information Resources Management Conference,* pp. 172-175.

Lally, L. & Garbushian, B. (2001). Hiring in the Post-Reengineering Environment: A Study Using Situationally Conditioned Belief. *Proceedings of the International Information Resources Management Conference,* pp. 234-237.

LaPorte, T.R. & Consolini, P.(1991). Working in Practice But Not in Theory. Theoretical Challenges of High Reliability Organizations. *Journal of Public Administration,* 1, 19-47.

Lefkon, D. (1997). Seven Work Plans for Year 2000 Upgrade Projects. *Communications of the ACM, 40*(5), 111-113.

Leveson, N. (1995). *Safeware: System Safety and Computers,* New York: Addison-Wesley Publishing Company.

Merian, L. (2001). Y2K Plans Aided in Recovery, But More Planning Needed. *Computerworld,* September, 19.

National Research Council. (2002). *Making the nation safer: The role of science and technology in countering terrorism.* Washington, D.C.: National Academies Press.

Nemeth, D. G., Creveling, C. C., Hearn, G.E., & Lambros, J.D. (1998). The Bray Y2K Survey: Positive Trends in Survey Data, *http:// www.year2000.com/archive/bray.html.*

Nolan, R. (1998). Connectivity and Control in the Year 2000 and Beyond. *Harvard Business Review,* July-August, 148-166.

Pauchant, T. & Mitroff, I. (1992). *Transforming the Crisis Prone Organization.* San Francisco: Jossey-Bass.

Perrow, C. (1984) *Normal Accidents: Living with High Risk Technologies,* New York: Basic Books.

Perrow, C. (1999) *Normal Accidents: Living with High Risk Technologies 2nd Edition,* New York, Basic Books.

Peterson, I. (1995). *Fatal Defect: Chasing Killer Computer Bugs,* New York: Times Books.

Poulson, K. (1999). The Y2K solution: Run for your life! http://www.wired.com/ wired/archive/6.08/y2k.html.

Report of the U.S. Senate Select Committee on Intelligence and U.S. House Permanent Select Committee on Intelligence, December, 2002, pp. xi-xv.

Sagan, Scott. (1993). *The Limits of Safety.* Princeton New Jersey: Princeton University Press.

Thibodeau, P. (2001). Businesses Eye Y2K Effort as Model for Terrorism Fight. *Computerworld,* October 2.

Ullman, E. (1999). *The Myth of Order Wired.* http://wired.com/archive/7.04/ y2k_pr.html.

Verton, D. (2002). IT Key to Antiterror Defenses at Nation's Sea Ports,"Computerworld, January 12.

Westland, J. C. (2000). Modeling the Incidence of Postrelease Errors in Software Information Systems Research, September, pp. 320-324.

Wielawski, I. (2003). Post 9/11 drills aid club fire rescue. *New York Times,* March 23.

Yourdon, E. (2002). *Byte Wars: The Impact of September 11 on Information Technology.* New Jersey: Prentice Hall.

## Chapter X

# Cryptography:
# Deciphering Its Progress

Leslie Leong
Central Connecticut State University, USA

Andrzej T. Jarmoszko
Central Connecticut State University, USA

## Abstract

*The emphasis and increased awareness of information security require an understanding and knowledge of the technology that supports it. As the risks and complexity of security keep growing along with the development of the Internet and e-commerce, securing information has become of utmost importance. With the worries about cyber-terrorism, hackers, and white-collar crimes, the demand for a stronger security mechanism in cryptography becomes apparent. Although wireless cryptography is still in its infancy, there are encryption technologies that may support the limitations of a mobile device.*

# Introduction

Protecting information has always been a great concern. Securing information can be accomplished through cryptography and an encoding process that is analogous to encryption. Both terms will be used interchangeably throughout this chapter. Cryptography is not a product of the modern technological age. In fact this technique has been known and applied for centuries to intelligently and mathematically scramble messages. Historical evidence reveals cryptography that dates back to ancient periods. In 1900 B.C. an Egyptian scribe used non-standard hieroglyphs in an inscription (Ellison, 2001). Julius Caesar's troops used it to communicate with each other from unsafe distances apart (Ackerman, 1999). The Caesar-shift substitution cipher is one of the easiest encryption techniques (Davis & Bennamati, 2003). Cryptography did not begin to advance into the powerful ciphering tool that it is today until the 20th Century. In the 1920s, cryptography and cryptanalysis became popular with governments (Ellison, 2001). During World War II, German forces used an encrypting machine called Enigma. It was constructed using mathematical and technological advances that were available at the time (it was cracked by three Polish mathematicians and provided valuable information to the Allied forces). Until the mid-1990s ciphering text was primarily a focus of national security, but with the emergence of electronic commerce, the private sector expanded its implementations. Typical Internet transactions require parties to exchange valuable personal and financial information via communications channels. To ensure protection of the exchanges, many companies turned to cryptography.

This chapter will focus on the structure of private-key and public-key encryption methods. It will look at specific examples of security validations, the legal aspects regarding export controls, and finally explore the emerging technologies in wireless cryptography.

# Cryptography

*"Cryptography, from the Greek kryptos ('hidden') and graphein ('to write'), is defined as the creation of systems to render a message unintelligible to unauthorized readers. Cryptanalysis, in contrast, is the practice of breaking codes, usually when the key is not known. Cryptology is the study of the two disciplines."* (Petras, 2001, p. 689)

The simplest forms of encrypting involve substituting message characters with a set of different characters that serve as identifiers to the characters of the message. Another method is switching the order of ASCII characters so they appear meaningless; for example, "ABC" becomes "BCD" (Boneh et al., 2001). Seemingly meaningless encrypted or cipher text is in actuality a mathematically organized set of characters that if used with an appropriate key, can be decrypted to its *plain text* state and thereby readable by the user. A key is a stream of bits of a set length created by a computer to encrypt or decrypt a message. Every cryptographic system consists of taking a message called *plain text,* applying a cipher or code associated with a particular key that enciphers or encodes the message, and producing *cipher text.* After the message is received, the rightful recipient, who possesses the key, can decipher or decode the message. Two widely used methods of cryptography are private-key (symmetric) and public-key (asymmetric) encryption. Both are highly popular in electronic communications and electronic transactions.

## Symmetric Encryption

Symmetric encryption, also known as private-key encryption or secret-key encryption, is the preferred method used by the government. Caesar-shift substitution ciphers fall into this category (Davis & Bennamati, 2003). In the event that two hosts want to safely exchange information via the Internet, they establish a connection using a single mutually agreed upon key called a private key. It is called private because it is not intended to be available to the public, but rather it is only to be shared by the two trusted parties. The two hosts establish a safe connection where each side represents a valid authentication of self-identity. Between each other, they exchange the private key that both will use to encrypt and decrypt the messages that they will be sending to one another. This private shared key will "lock" a *plain text* message into a cipher text message immediately before sending it from the source host; it will "unlock" the cipher text into a plain text message at the destination host.

An example of symmetric cryptography is the Data Encryption Standard (DES), a 56-bits key adopted in 1976 by the National Institute of Standards and Technology (NIST) as a federal standard for encryption of non-classified information. In 1997, it was realized that a 56-bit key was insufficient to maintain an acceptable level of secure communications. In 1998, the Electronic Frontier Foundation (EFF, 1998) broke DES in less than three days. Today the

U.S. government uses the Triple DES, which is a successor of DES. Triple DES is an advanced application of the principles of DES and is designed to replace DES. The main difference between the two is the size of the private key. 3DES employs 112- and 128-bit keys. In general, the difficulty in cracking a key is directly proportional to the size of the key. 3DES employs 48 rounds in its encryption computation, generating ciphers that are approximately $2^{56}$ stronger than standard DES while requiring only three times longer to process. The larger the key, the more processing time is needed to decode it, but the benefits of a secure communication far outweigh the processing time. Nevertheless, any current encryption system like 3DES or RSA will become obsolete as the world advances into faster and more powerful computers.

## Asymmetric Encryption

Asymmetric encryption, also known as public-key encryption, uses a public-and-private key combination. This type of cryptography is most widely represented by a system called RSA (Rivest-Shamir-Adelman), named after the three M.I.T. professors who developed it in 1977. "RSA is one of the most popular public-key cryptosystems in use in the world today" (Petras, 2001, p. 691). The technique of asymmetric encryption is more complex compared to symmetric encryption. It is also called a public-key encryption because it involves using a public key that is available to anyone on the Internet who wishes to establish a connection, also called a session, with the owner of that public key. When a server wants to make a connection with a client browser, it sends an initial message to communicate its system requirements for a connection. The browser asks for digital certification from the server to prove the identity of the server, in a process called authentication. The server then sends its identity as well as its public key in the form of a digital certificate signed by a legitimate certification authority.

After a successful authentication, the browser generates a private key that will be shared by the two parties for the duration of the session, or connection. This private key is also called a session key. It is actually a second private key held by this computer. Each computer that uses software equipped with asymmetric encryption such as Netscape Communicator and Microsoft Internet Explorer browsers, operating systems, applications like Microsoft Word, and e-mail holds a permanent (until one chooses to alter it) private key, most commonly inside the hard drive, that is not revealed to anyone by its owner. The session key is a different kind of private key designed for a one-session use. The browser generates this session key and encrypts it with the server's public key that was delivered by the server in its digital certificate. Once encrypted, the session key can now be safely sent via the Internet to the server. The server receives the key and decrypts it using its own secret, undisclosed, permanent private key. Even though this permanent private key is never exchanged it is used to decrypt its corresponding public key's encryptions. Once the packet is decrypted, the shared session key is revealed to the server. From this point on the two parties engage in private-key (session key) encryption for the remainder of the session/connection. Until this session is closed, for example until the browser leaves the server's Website, the parties encrypt and decrypt their exchanges using a single, shared session key. It is called a session key because when the session ends this session key is discarded, and never used again. The moment the two hosts reestablish a connection the asymmetric process begins all over again, and a new temporary private key, meaning a session key, is generated.

# Pros and Cons of Cryptography Techniques

Both symmetric and asymmetric encryption methods have their comparable strengths and weaknesses. One problem that arises regarding private-key ciphering is how to safely send the shared private key to the party that is requesting a connection. A private key can neither encrypt nor decrypt itself. Sending the private key in plain text via the Internet defeats the purpose of cryptography altogether. The two parties might as well abandon encrypting because once a hacker gets a hold of a private key, he or she can easily read any encrypted information sent, for example names, addresses, passwords, userIDs, credit card numbers, and so forth. The two parties need to negotiate how to safely exchange the shared private key such as by postal mail.

Despite this obstacle, companies engage in symmetric encryption because it is done much faster than asymmetric encryption (often by a difference of 45 seconds). One of the reasons for this speed differential is that symmetric encryption involves much less processing, and no public key encryption, nor generating of session key with each connection. In addition, PKI (Public Key Infrastructure) uses much larger numbers than private-key encryption. Whereas symmetric keys are typically short (40 or 128 bits), asymmetric key technology requires the keys to be long (over 1,000 bits), which makes sending and receiving long messages a timely process (Hillison et al., 2001). RSA protection is based on the difficulty of factoring a number $n$. This number $n$ is a product of two prime numbers $p$ and $q$, with each being 100-digits long, making the number $n$ approximately 200-digits long. For a hacker to crack by factoring the 200-digits long number $n$, with today's most powerful computers put together, it would take approximately 4 billion years (Petras, 2001). In essence, public-key encryption uses large numbers that increase encryption-processing time. There are several standards for content encryption over the Internet. IPSEC, PGP, SSL, TLS— all these serve for adequate security, if used with keys that are cryptographically strong and generated and handled in a secure manner (Posch, 2001).

A disadvantage to private-key ciphering is that both parties have access to their private key. A company puts itself at risk by sharing a private key with another. This broadens the spectrum for hacker access links to weak parts of the network infrastructures of the connected parties. Companies that establish safe connections use a separate private key for each one of their Internet exchange partners. It would be unwise to use the same private key within the symmetric infrastructure with several hosts. Therefore, companies need to keep in their secure hard drive one private key per partner, resulting in a lot of private keys to consider in a large environment such as the Internet.

This problem is resolved in the PKI. The companies use their same permanent public and private keys with all the connecting hosts, and the session key is generated only temporarily for the duration of one session. As a result, the session key does not need to be stored. Another benefit of a session key is that if a hacker somehow taps into the system and cracks the session key, he or she will only be able to access the contents of the packet captured for this one particular session/connection. The moment the companies break the connection, they discard the captured session key. With each new connection they generate a new session key, forcing the hacker to start all over again the process of cracking the encrypting/decrypting key. If the hacker does capture a bunch of packets for the cracked session key, he/she may still not get enough information out of this fraction of the complete message that was sent.

Asymmetric encryption offers the advantage that private keys are unique to an entity (user, server, and so on), and they can be used to create digital signatures (Boneh et al., 2001). Consequently, they are very safe guards against the breach of secrecy. A unique property of the RSA cryptosystem is that a private key cannot encrypt/decrypt a private key, and a public key cannot encrypt/decrypt a public key. Instead, a private-key is used to encrypt/decrypt a public key, and a public key is used to encrypt/decrypt a private key.

An inherent problem with asymmetric encryption is that since the public keys are widely available to the public on the Internet, anyone can create a fake public key and assume a fake representation or identity by it. A solution to unauthorized public keys is a certificate authority (CA), such as VeriSign. This firm specializes in verifying the identity of a host, or a Web-based product-manufacturing company, or anyone involved in e-commerce. VeriSign signs a digital certificate with its own key to anyone who provides legitimate identification proof. VeriSign then sends the applicant the certificate, which subsequently proves that the owner of it is the rightful owner of the Website, product, or service that it wants to provide. When someone performs verification that the partner who sent this certificate to him or her for authentication is the rightful, registered owner of it, that individual's browser runs through its installed list of CAs and looks for the public key of the CA that corresponds with the public key on the digital certificate. If they match, the authentication is completed positively.

# Encryption and Law

The United States has been a leader in the development and implementation of information security legislation to prevent misuse and exploitation of information and information technology. The protection of national security, trade secrets, and a variety of other state and private assets has led to several laws impacting what information and information management and security resources may be exported from the United States. These laws attempt to stem the theft of information by establishing strong penalties for related crimes. The Economic Espionage Act of 1996 and the Security and Freedom Through Encryption Act of 1999 are two laws that directly affect information security.

Encryption imposes several legal challenges brought by the U.S. government. The government has put in effect encryption export laws with intent to inhibit exporting encryption products to foreign countries. The government fears that wide distribution of American intelligence will enable the foreign powers, as well as the criminal entities, to conduct safe communication across the border.

In an attempt to protect American ingenuity, intellectual property, and competitive advantage, Congress passed the Economic Espionage Act (EEA) in 1996. This law attempts to prevent trade secrets from being illegally shared. On the other hand, the private sector opposes such laws on the basis of the First Amendment, which provides freedom of speech. Encryption is tolerated as a form of expression and therefore it has limitations on the export laws. The export laws, however, continue to be in effect, attacking individual aspects of constitutionally protected materials.

The Security and Freedom Through Encryption Act of 1999 was an attempt by Congress to provide guidance on the use of encryption, and provided measures of public protection from government intervention. The act's provision include the following:

- Reinforces an individual's right to use or sell encryption algorithms, without concern for regulations requiring some form of key registration. Key registration is the storage of a cryptographic key with another party to be used to break the encryption data. This is often called "key escrow".

- Prohibits the federal government from requiring the use of encryption for contracts, grants, and other official documents, and correspondences.

- States that the use of encryption is not probable cause to suspect criminal activity.

- Relaxes export restrictions by amending the Export Administration Act of 1979.

- Provides additional penalties for the use of encryption in the commission of a criminal act. (Source: http://www.bxa.doc.gov)


"On June 30, 2000, President Clinton signed into law the Electronic Signatures in Global and National Commerce Act (E-Signature Act). The law took effect on October 1, 2000. The act allows electronic signatures or documents to satisfy most existing legal requirements for written signatures, disclosures, or records. It does not, however, mean that all e-signatures and records are now automatically legally binding" (Hillison et al., 2001, p. 4). The new law makes it clear that certain transactions have to be performed in writing and accompanied by a handwritten signature. These include: wills, codicils, testamentary trusts, cancellation notices involving health and life insurance (other than annuities), family law documents (e.g., divorce decrees), court orders and notices, and default notices and foreclosure documents related to a person's primary residence, and records and documents governed by the Uniform Commercial Code.

# Wireless Cryptography

As mobile devices become more integrated into corporate networks and as they are used to fulfill identification and authorization functions for their owners, security becomes of increasing importance. While it is the applications and systems operations levels that will ultimately determine the effectiveness of how secure a particular system is, there are a variety of ways in which the operating system can support these high-level developments.

Schemes to support authentication for authorization and identification such as password, finger and palm print ID, and retina scan, among others, are used in addition to the Secure Socket Layer (SSL) security in the wireless environment. The other area of concern with wireless networking is the vulnerability to interception of wireless exchanges. As a result, the need for encryption that supports the wireless transaction and communication must be addressed.

The booming wireless devices in communications and electronic transactions require an encryption that requires a new method of calculation and processing capabilities. Due to the limitations that a wireless device can support, computational efficiency and storage capacity needs to be addressed. The Wired Equivalent Privacy (WEP) technology has failed to provide the necessary security and encryption on a wireless connection (Mehta, 2001; NetworkWorld, 2002). Therefore, hashing algorithms and mathematical methods may be a viable encryption method for mobile devices.

The Secure Hash Algorithm (SHA), developed by the National Institute of Standards and Technology (NIST), along with the NSA for use with the Digital Signature Standard (DSS) is specified within the Secure Hash Standard (SHS). The Secure Hash Algorithm takes a message of less than $2^{64}$ bits in length and produces a 160-bit message digest, which is designed so that it should be computationally expensive to find a text that matches a given hash. Message digest, also called one-way hashing function, is a function that calculates the "fingerprint" of a message. Two different messages, no matter how minor the differences are, are ensured to yield two different fingerprints (or digests) by the one-way hashing function. As of today, there is no known report of the breaking of SHA-1.

The Secure Hashing Algorithm (SHA-1) summarizes the message to be transmitted. This method reduces computational time for private key signing and public key authentication, requires less bandwidth and the compressed message is faster to encrypt.

Mathematical routines that best fit the wireless environment are used by the Elliptic Curve Cryptosystem (ECC). Rather than using linear algebra for public key and for digital signature generation, ECC systems use the Discrete Logarithm Problem (DLP). DLP uses a formula with the groups of points on the elliptical curve and determines the private/public key relationship. This mathematical method of encryption uses a technique called the Elliptic Curve Discrete Logarithm Problem (ECDLP). Leading cryptographers concluded that the ECDLP in fact requires fully exponential time to resolve. Elliptic Curve Cryptosystem is recognized as a mature technology and companies are now implementing it for widespread deployment (Certicom, 2000).

In the wireless environment, TeleHubLink Corporation is finalizing the development of its ASIC prototype microchip for the wireless communications market. IBM assisted with the design and prototyping of TeleHubLink Hornet cryptographic ASICs. TeleHubLink's Wireless Encryption Technology Division is developing a family of secure wireless and broadband communications ASICs under the trademarked name of Hornet that use a series of powerful and high-speed stream ciphers in cost-sensitive markets like telephony as well as the wireless communication technologies (Fabtech, 2001).

IEEE 802.11b or, more commonly, wireless Ethernet, has a few basic security features that aim to make communications as secure as the wired Ethernet equivalent. However, with a wireless network things are a little more complicated. Signals are broadcast in all directions from both mobile terminals and access points. All an intruder needs to do is to get in range equipped with a laptop kitted out with a suitable wireless NIC. To protect IEEE 802.11b networks from this form of intrusion, the standard specifies an encryption mechanism called Wired Equivalent Privacy (WEP). WEP uses the Rivest Cipher #4 (RC4) encryption algorithm. RC4 is also used by the Secure Socket Layer (SSL) in Web browsers. However, the particular way in which it has been implemented in WEP makes the network extremely vulnerable.

For these several reasons, organizations have been reluctant to use a wireless network. Security is being improved and current systems can be made much more secure than they are "out of the box" (Gast, 2002). Currently, the new standard, IEEE 802.11g™ Standard Extends Data Rate of IEEE 802.11b™ WLANs to 54 Mbps from 11 Mbps (Kerry & McCabe, 2003). However, the new standard was compatible with the equipment already in the field. The new product, although apparently more secure and perhaps less subject to interference, is also more expensive and may have a shorter range, thus requiring more units.

# Conclusion

As long as there is a need for secure communication, there will be some form of cryptography method applied. With the advancement of the technological age an ordinary user will become less and less responsible for knowing the details of his or her system. For example, now cryptography takes place automatically without the need to involve the user unless he or she wishes to custom design his or her system. The communication between the network interface card (NIC) and routers is an independent activity that frees the users from direct involvement in cryptography.

The sole purpose of technology is to support our daily functions in the most efficient and effective way. It is important to keep in mind that the word "secure" does not mirror its definition in the world of constant technological advancements. The core basis for the strength of any encryption method is not the logic of a mathematical algorithm. The trick to render a hacker unsuccessful will always be the amount of time it will take to crack the code. Understanding the functionality of an algorithm can be achieved, but equipping oneself with the powerful resources is the greatest challenge to an intelligent hacker. A faster computer of tomorrow may be simply calculating more functions of an algorithm than a faster computer of today using the same, even known algorithm. Reducing the time to decrypt programs such as 3DES is an enormously resource-demanding venture.

Modern cryptographers are constantly working on new algorithms that are exceedingly difficult to break. Some of these developments are in quantum computing, which speeds up the calculations, and DNA cryptography, where DNA molecules are viewed as computers capable of rapid calculation. Other current work includes advances in elliptic curves cryptography (ECC), hyper-elliptic cryptosystems, RSA variants and optimizations, multivariate quadratic equations over a finite field (the MQ problem) and lattices. When coupled with other solid methods of access control these new technological advancements may reduce hackers and white-collar crime, but loopholes will still need to be addressed.

As in any product life cycle, some cryptosystems will have very short lives (such as the case with DES) and others may span a century. In cryptography, this happens when an easily exploitable flaw is found in the algorithm and the underlying cryptosystem is deemed beyond repair. The efficacy of these products remains generally unknowable by the people who buy and implement them before testing. Implementations were in the field before the protocol weaknesses were fully understood.

In securing wireless communication, more research and testing need to be conducted before the wireless technology can be implemented. Most business functions have a financial justification as the basis for investments in the technologies that support them. Replacing and supporting an equipment to support a replacement for DES, for example, cannot happen quickly. The new standard of IEEE 802.11g boasts a higher data transfer rate, but organizations may wish to wait until further testing is conducted and the technology matures before engaging in sizable financial investment. Needless to say, the security in wireless technology is still in its infancy.

In the legal aspect, export control must be tightened. Advances in cryptography cannot be simply made available in the hands of those engaging in criminal activities. It is appropriate to argue that export controls provide no control over the travel of informational products and services. Any person can walk into an American store, purchase a software technology and find a means of transporting it abroad. "The export controls have been ineffective and counterproductive policy and are arguably unconstitutional under the First Amendment. However, export controls are the only viable solution to the intelligence gathering problem and will need to survive these political and legal attacks or our national security could be jeopardized" (Ackerman, 1999, p. 17).

# References

Ackerman, R. (1999). Digital handshake characterizes defense e-commerce. *National Security,* March, 17. Available: http://www.us.net/signal/subjectindex/nationalsec.html

Boneh, D., Durfee, G., & Franklin, M. (2001). Lower bounds for multicast message authentication. *Advances in cryptology.* Eurocrypt.

Certicom. (2000, July). *The elliptic curve cryptosystem* [Online]. Available:*http://www.certicom.com/resources/download/EccWhite3. pdf.*

Davis, W., & Bennamati, J. (2003). *E-commerce basics: Technology foundations and e-business applications.* New York: Addison-Wesley Press.

EFF. (1998.) "EFF DES Cracker" machine brings honesty to crypto debate. Electronic Frontier Foundation proves that DES is not secure [Online]. Available:*http://www.eff.org/Privacy/Crypto_misc/DESCracker/ HTML/19980716_eff_descracker_pressrel.html.*

Ellison, C. (2001). Cryptography timeline [Online]. Available:http:// www.math.nmsu.edu/crypto/public_html/Timeline.html.

Fabtech. (2001). *IBM schedules TeleHubLink's wireless encryption microchip for manufacturing* [Online]. Available: http://www.semiconductor fabtech.com/site-global/news/2001/05/30.01.shtml.

Gast, M.S. (2002). *802.11 wireless networks: The differences guide.* ISBN:0596001835. UK: O'Reilly.

Hillison, W., Pacini, C., & Sinason, D. (2001). Electronic signatures and encryption. *The CPA Journal*, August, 20-25.

Kerry, S., & McCabe, K. (2003). *Popular wireless local area networks gain large boost in speed* [Online]. Available: http://standards.ieee.org/ announcements/80211gfinal.html.

Mehta, P.C. (2001). Wired equivalent privacy vulnerability. Sans Institute. Available: *http://rr.sans.org/wireless/equiv.php.*

*Network World. (2002, November).* WEP (Wired Equivalent Privacy) *[Online]. Available:*http://napps.nwfusion.com/links/Encyclopedia/W/ 715.html.

Petras, R.T. (2001). Privacy for the Twenty-First Century: Cryptography. *Mathematics Teacher, 94*(8), 689-691.

Posch, R. (2001). Will Internet ever be secured? *Journal of Universal Computer Science, 7*(5), 447-456.

Chapter XI

# A Method of Assessing Information System Security Controls

Malcolm R. Pattinson
University of South Australia, Australia

## Abstract

*This chapter introduces a method of assessing the state of an organization's information system security by evaluating the effectiveness of the various IS controls that are in place. It describes how the Goal Attainment Scaling (GAS) methodology (Kiresuk, Smith & Cardillo, 1994) was used within a South Australian Government Agency and summarises the results of this research. The major purpose of this research was to investigate whether the GAS methodology is a feasible method of assessing the state of security of an organization's information systems. Additional objectives of this research were to determine the suitability of the GAS methodology as a self-evaluation tool and its usefulness in determining the extent of compliance with a mandated IS security standard.*

# Introduction

Information System (IS) security has become a critical issue for most organizations today, particularly those that have a large investment in information technology (IT). However, management and internal auditors still ask, "Is our information secure?" or "Do we have the necessary blend of controls in place to withstand the various threats to our information?" These questions are still very difficult to answer and in the past, management has sought answers by conducting computer security reviews and risk analyses by both internal and external security specialists. These projects are typically expensive, time-consuming and resource-intensive because they were originally designed for use in large organizations with mainframe computers. With the advent of client-server technology and distributed computing via networks there is a need for a self-assessment technique or measuring device that is inexpensive and easy-to-use whilst providing management with a reliable indication of just how effective their IS security controls are.

This chapter describes a program evaluation methodology (Isaac & Michael, 1995; Owen, 1993) known as Goal Attainment Scaling (GAS) (Kiresuk, Smith & Cardillo, 1994), which has been used predominantly in the disciplines of health, social work and education and applies it within the discipline of IS security. Whilst traditionally the GAS methodology has been used to detect positive or negative changes in a patient's mental health, this application of GAS attempts to assess the changes in the "health and safety" of a computer system over time. More specifically, this chapter describes how the GAS methodology was used to evaluate IS security within a South Australian government agency by assessing the state, condition and quality of IS security controls in place.

# Assessing Information System Security Controls

An IS security control can take many forms. It can be a hardware device or a computer program or indeed a management process. In all cases its purpose is to prevent, avoid, detect or prepare for breaches of security that threaten the confidentiality, integrity or availability of information processed by computer systems. But how can the effectiveness of such a control be measured? The Australian Standard for Risk Management (Standards Australia/Standards New Zealand, 1999) defines the effectiveness of a control as a measure of how well the control objectives (and/or risk objectives) are met (i.e., how the risk of not achieving these objectives is reduced). For example, a typical control/ risk objective is to prevent, or at least minimise, security breaches. In this case, it could be argued that the true effectiveness of a control of this nature is a measure of the number of threats that occur in spite of its presence. In other words, a control's effectiveness is a measure of the number of security breaches that it prevents. For example, if no security breaches occur, then it could be claimed that the controls that relate to that threat are 100% effective. However, there are two problems with this argument. Firstly, there is rarely a one-to-one relationship between a threat and a control. To explain this, consider the threat of unauthorised access to sensitive information. This is controlled by a number of physical controls such as locks, security guards and alarm systems and also by a number of logical controls such as passwords, encryption and authentication software. As such, it is difficult to isolate any of these controls and establish its sole effectiveness in preventing unauthorised access.

The second problem with attempting to assess the effectiveness of a control by measuring the number of security breaches it prevents relates to the fortuitous nature of threats. The occurrence of some threats is often unpredictable and independent of the controls in place. For example, some premises without locks may never have a break-in, whilst others with numerous locks may have two or three each year. There is a degree of luck involved. Therefore, the fact that no security breaches occur may be just good fortune and not a measure of how many breaches were prevented or a measure of the effectiveness of the controls in place.

The approach described in this chapter concentrates on assessing or evaluating the state, quality or condition of the controls that are in place. It does *not* assess the effectiveness of controls by measuring the number of security breaches that have occurred or have been prevented. As stated earlier, this "quality" or "condition" can relate to a number of characteristics of the control in question. For example, a control can be measured in terms of how well it has been implemented, that is, its "implemented-ness". As an analogy, compare this to assessing the effectiveness of a raincoat in preventing a person from getting wet. The obvious approach would be to measure how wet the wearer gets when it rains. Alternatively, one could get an indication of its effectiveness by assessing the raincoat in terms of its quality, that is, how well is it made? If it is designed well and made from high quality materials it is likely to be effective. Another option could be to assess the raincoat in terms of how well it is being applied, that is, is it being used properly? This may also give an indication of how effective it is. The point being made here is that the effectiveness of a raincoat, or indeed a management control, can be measured in more than one way.

# What is Goal Attainment Scaling (GAS)?

The GAS methodology is a program evaluation methodology used to evaluate the effectiveness of a program or project (Kiresuk et al., 1994). A program evaluation methodology is a process of determining how well a particular program is achieving or has achieved its stated aims and objectives. Kiresuk and Lund (1982) state that program evaluation is a process of establishing "...the degree to which an organization is doing what it is supposed to do and achieving what it is supposed to achieve" (p. 227).

One of the essential components of the GAS methodology is the evaluation instrument. This is primarily a table or matrix whereby the columns represent objectives to be assessed and the rows represent levels of attainment of those objectives. Kiresuk et al. (1994) refer to these objectives as goals or scales within a GAS Follow-up Guide. The rows represent contiguous descriptions of the degree of expected goal outcome. These can range from the best-case level of goal attainment to the worst case, with the middle row being the most likely level of goal attainment. The sample GAS Follow-up Guide in Figure 1 is a portion of one of seven follow-up guides that comprised the complete evaluation tool used in the case study described herein.

The GAS goals that are being evaluated are, in fact, IS controls. This is an important concept because these types of "operational" GAS goals can be evaluated as either an outcome or a process (refer to "Evaluating Processes or Outcomes" in the Discussion section).

*Figure 1. Sample GAS follow-up guide*

**DATA CONFIDENTIALITY**
Objective: To ensure that only authorised people have access to classified information.

| LEVEL OF ATTAINMENT OF IS CONTROL | CLEAN DESK POLICY | CONTROL THE ENTRY AND SUPERVISION OF SERVICE PERSONNEL FROM EXTERNAL ORGANISATIONS | PHYSICAL ACCESS CONTROLS FOR OFF-SITE REMOVABLE STORAGE MEDIA |
|---|---|---|---|
| Much more than Acceptable level | Diskettes are stored in cabinets when they are not being used. Sensitive data media are stored in locked fire-resistant cabinets when not required & desks are cleared of all papers when desk is vacated. | Service people do NOT have easy physical access to any IT equipment (incl. workstations) & they must wear identity badges. All entry is logged and log is periodically audited. Also, they are escorted at all times. | Stored in fireproof, lockable cabinets. Physical access is restricted to authorised personnel only. Approval required to remove. Transported by professional security organisation. |
| Somewhat more than Acceptable level | Diskettes are stored in cabinets when they are not being used. Sensitive data media are stored in locked fire-resistant cabinets when not required. | Service people do NOT have easy physical access to any IT equipment (incl. workstations) & they must wear identity badges. All entry is logged and log is periodically audited. | Stored in fireproof, lockable cabinets. Physical access is restricted to authorised personnel only. Approval required to remove. |
| Acceptable level | Diskettes are stored in cabinets when they are not being used. | Service people do NOT have easy physical access to any IT equipment (incl. workstations). | Stored in fireproof, lockable cabinets. Physical access is restricted to authorised personnel only. |
| Somewhat less than Acceptable level | Diskettes are stored in a lockable diskette box on the desk or similar when not being used. | Service people have easy physical access to workstations but NOT network server(s) and comms. equipment. | Stored in fireproof, lockable cabinets but physical access is NOT strictly controlled. |
| Much less than Acceptable level | In general, desks are NOT cleared of papers and diskettes when the desk is vacated. | Service people have easy physical access to ALL IT equipment (incl. workstations). | NOT stored in fireproof, lockable cabinets and physical access is NOT strictly controlled. |

# How GAS was used in a Case Study

The case study organization referred to in this chapter was a South Australian government agency whose major function is to manage deceased estates. It has a total staff of 120. The IT department consists of four people, one of whom is the IT Manager, who is responsible for IS security within the agency. The IT facilities consist of a 150-workstation local area network (LAN) bridged to a mid-range computer system that runs their major Trust Accounting system. The hardware facilities and the software development function have been outsourced to various third party organizations. Although this agency is considered "small" relative to other South Australian government agencies, the issue of IS security is perhaps more important than with most other agencies because of the due diligence associated with the management of funds held in trust.

This case study comprised three phases, namely:

1.   Develop the GAS evaluation instrument.
2.   Use the GAS evaluation instrument.
3.   Analyse the evaluation results and report to management.

## Phase 1: Develop the GAS Evaluation Instrument

The GAS evaluation instrument developed in this phase of the case study was to represent all aspects of IS security for an organization. Because IS security is such a wide domain, the instrument needed more than a single follow-up guide. Consequently, it was necessary to develop a set of GAS follow-up guides, one for each category of IS security. The resulting GAS evaluation instrument, comprising seven follow-up guides, was developed in accordance with the nine-step process (Kiresuk et al., 1994, pp. 7-9). The titles for each of the steps are not identical to the titles of the Kiresuk et al. (1994) method, but have been modified to suit the IS security environment and in accordance with the case study agency's preference. Furthermore, step nine became redundant since all controls were addressed in each step. The nine steps are as follows:

1.   Identify the security areas to be focused on.

2.   Translate the selected security areas into several controls.

3.   Choose a brief title for each control.

4.   Select an indicator for each control.

5.   Specify the minimal acceptable level of attainment for each control.

6.   Review the acceptable levels of attainment.

7.   Specify the "somewhat more" and "somewhat less" than acceptable levels of attainment.

8.   Specify the "much more" and "much less" than acceptable levels of attainment.

9.   Repeat these scaling steps for each of the controls in each follow-up guide.

Each of these Phase 1 steps is described in detail below:

### *Step 1: Identify the security areas to be focused on.*

For the purpose of this research, the issues to be focused on included all aspects of IS security within an organization. It would be quite acceptable to focus on a specific category of IS security, but the case study agency agreed to address the whole IS security spectrum. IS security is a wide domain and it was necessary to define categories that were a manageable size. Rather than "re-invent the wheel," all that was necessary was to refer to any number of published standards on computer security. For example, the Standard AS/NZS ISO/IEC 17799 (Standards Australia/Standards New Zealand, 2001) breaks down IS security into 10 categories and 36 sub-categories. This case study adopted a mandated government standard comprising seven IS security categories defined by the South Australian Government Information Technology Security Standards document (South Australian Government, 1994b) as follows:

• The Organization
• Authentication
• Access Control

- Data Integrity
- Data Confidentiality
- Data Availability
- Audit

## Step 2: Translate the selected security areas into several controls.

For each IS security category, the project team selected a set of IS controls that were considered most relevant, important and suitable for evaluation. Each of the selected IS controls would eventually be represented by a column in a GAS follow-up guide. The eventual number of IS controls selected ranged between 3 and 10 per follow-up guide, giving a total of 39.

This task of selecting an appropriate set of IS controls is critical to the success of using the GAS methodology. With this in mind, it is advisable that this evaluation methodology be complemented by an independent audit of all controls, particularly those compliance-driven controls that are excluded from the GAS follow-up guides because of their difficulty to scale.

## Step 3: Choose a brief title for each control.

This step required that an abbreviated title be devised for each IS control selected in the previous step so that the eventual user of the GAS follow-up guides would easily recognise the control being evaluated.

## Step 4: Select an indicator for each control.

This step relates to the criteria used to measure an IS control. The indicator is an element of measurement chosen to indicate the level of attainment of the IS control. For example, one IS control was "Establish formal policy with respect to software licences and use of unauthorised software". A possible indicator in determining the level of attainment of this control is the amount or extent of progress made towards developing such a document (e.g., fully developed, partly developed or not started).  An alternative indicator could be the frequency of occurrence of the IS control. For example, the control "Regular reviews of software and data content of critical systems" could be measured by the frequency with which a review is conducted (e.g., annually or every six months).

## Step 5: Specify the minimal acceptable level of attainment for each control.

This step required the development of narrative for the "zero" or middle cell for each control within each follow-up guide. This description represents the *minimum* level of acceptance (by management) of an IS control compared to the traditional "expected level of outcome" as per the Kiresuk et al. (1994) GAS methodology.

## Step 6: Review the acceptable levels of attainment.

The objective of this step is to confirm the relevance and understandability of the descriptions by potential users. It is also important that these descriptions represented minimal acceptable levels of attainment of IS controls by agency management.

## Step 7: Specify the "somewhat more" and "somewhat less" than acceptable levels of attainment.

This step required that narrative be developed that described the "somewhat more" and "somewhat less" than the minimum level of acceptance scenarios for each IS control.

## Step 8: Specify the "much more" and "much less" than acceptable levels of attainment.

This step required that narrative be developed, as in Step 7, but which described the "much more" and "much less" than the minimum level of acceptance scenarios for each IS control.

*Step 9: Repeat these scaling steps for each of the controls in each follow-up guide.*

Redundant.

## Phase 2: Use the GAS Evaluation Instrument

This phase involved the use of the individualised GAS instrument by conducting a pre-treatment evaluation of IS security within the agency followed by a post-treatment evaluation 15 months later using the same GAS instrument without modification. The reference to "treatment" in this case study was the implementation and maintenance of IS controls in response to the first evaluation. This phase formed a very small component of this whole research because the GAS methodology requires that most of the time and effort is spent on developing the instrument so that it is quick and easy to use.

The ease and speed with which this data collection process was conducted on both occasions is not only testimony to the efficiency of the GAS methodology, but satisfies Love's (1991) essential characteristics of a Rapid Assessment Instrument (RAI) when used in a self study situation. He states, "They must be short, easy to read, easy to complete, easy to understand, easy to score, and easy to interpret" (p. 106).

## Phase 3: Analyse the Evaluation Results and Report to Management

This phase involved the analysis of the data collected during the second phase. Raw scores were converted into GAS T-scores for each of the seven follow-up guides in accordance with the Kiresuk et al. (1994) methodology.

A GAS T-score is a linear transformation of the average of the raw scores in each follow-up guide using the formula documented by Kiresuk et al. (1994) and presented below:

$$T-score = 50 + \frac{10 \sum w_i x_i}{\sqrt{(1-p) \sum w_i^2 + p(\sum w_i)^2}}$$

where $x_i$ is the outcome score for the $i$th scale with a weight of $w_i$, and p is the weighted average inter-correlation of the scale scores and commonly set at 0.3. Scores on the individual scales between -2 and +2 each are assumed to have a theoretical distribution with a mean of zero and a standard deviation of 1. This formula then produces T-scores with a mean of 50 and a standard deviation of 10 when each scaled control is scored using the -2 to +2 scale.

For both the pre-treatment and the post-treatment evaluations, two T-scores were calculated for each follow-up guide, a non-weighted T-score and a weighted T-score. GAS T-scores for weighted controls were generated for the purpose of seeing what difference weighting actually made. GAS T-scores were then compared, interpreted and written up in business report form for management of the agency.

Figure 2 shows the non-weighted GAS T-scores for the pre-treatment and post-treatment evaluations. The "treatment" here is the action taken (or not taken) by management in response to the pre-treatment evaluation results. Figure 3 shows the weighted GAS T-scores for the pre-treatment and post-treatment evaluations. Both charts highlight the negative change in the rating of the level of security over the 15 months between evaluations.

*Figure 2. Non-weighted GAS T-scores*

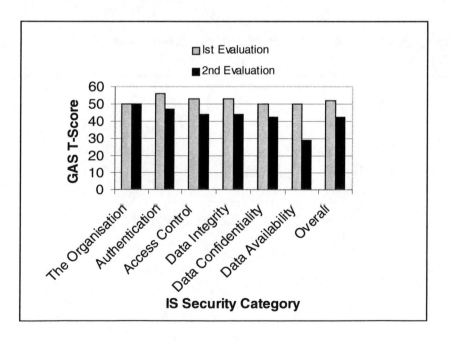

*Figure 3. Weighted GAS T-scores*

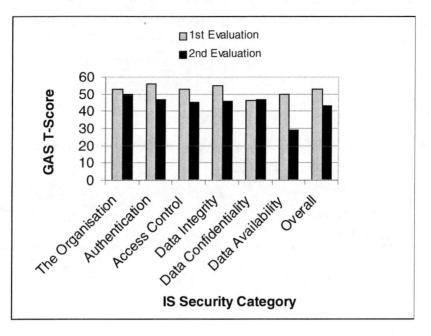

A GAS T-score of 50 or more indicates that, on average, the controls specified within a GAS follow-up guide for a particular IS security category are considered acceptable to management. If these controls are representative of all controls specified for this IS security category within the adopted standard, then it can be argued that the controls in place are generally compliant with the standard. The extent of compliance is reflected in the amount by which the score is greater than 50. Conversely, a GAS T-score of less than 50 for a particular IS security category indicates that, in general, the IS controls in place are *not* acceptable to management and therefore (it could be argued) they are not compliant with the adopted standard.

In the first or pre-treatment evaluation the non-weighted GAS T-scores were all greater than or equal to 50. This means that, on average, the controls within each IS security category are considered to be acceptable relative to the government standards on which they are based. That is, on average across the scaled data cells, the IS controls are considered compliant with those standards provided the levels of attainment have been set correctly.

The second or post-treatment evaluation, conducted 15 months after the first, showed a significant decline in the GAS T-scores for each of the IS security categories and for overall security, using both weighted and non-weighted controls. In particular, the IS category "Data Availability" showed a dramatic decline in the supposed level of security since the first evaluation.

In summary, the overall results indicated that the level of IS security within the agency had dropped between the two evaluations in all IS categories except one, which showed no change. In fact, the overall results shown in Figures 2 and 3 for both weighted and non-weighted GAS T-scores indicate a substantial negative change in the state of IS controls between the first and second evaluations.

# Discussion

This section discusses the following major issues associated with this research:

- The nature of information system security
- Using a baseline standard to select controls
- Evaluating processes or outcomes
- The need for an independent audit
- The need for GAS training
- Evaluating a non-human service
- Criticisms of GAS

## The Nature of Information System Security

Information system security relates to the adequacy of management controls to prevent, avoid, detect, deter and recover from a whole range of threats that could cause damage or disruption to computer systems. Total security is rarely achieved because it requires that the risk of every possible threat is zero, and for this to be the case, the computer system would not be functional. This is impractical and a balance needs to be struck between the impact that the threat will have on the business and the cost of the controls necessary to minimise risks. Therefore the aim of management should be to reduce the risks to an acceptable level and not to eliminate them. But what is an acceptable level of risk? This is a complex question and the answer is different for each organization, depending on a number of factors such as management's propensity to take risks and the sensitivity of the information stored and processed. These risk objectives should be established as a result of conducting a risk analysis before controls are selected and implemented. The methodology presented herein does not purport to measure or assess risk, but instead, it evaluates the condition or state or quality of IS controls that have been put in place. Consequently, the methodology does not attempt to determine the extent to which the controls in place achieve the risk objectives of the organization. It does, however, attempt to indicate the level of IS security relative to acceptable IS control benchmarks that are set by management as part of the methodology. This assumes that these acceptable levels of IS controls are in tune with the organization's risk objectives.

The GAS methodology outlined herein gives an indication only of the state of security of each individual IS security category and the state of overall IS security by the aggregation of all categories. If an IS security category receives a very high score, this does not necessarily mean that all controls within the category are adequate. Similarly, if overall IS security is considered adequate, this does not mean that there are not areas of weakness that need attention. This "weakest link in the chain" argument is particularly relevant in relation to security controls in general and is a possible criticism of the methodology.

## Using a Baseline Standard to Select Controls

When the GAS methodology is used in collaboration with a set of IS security standards, the goal setting process described by Kiresuk et al. (1994) is assisted by the existence of a complete set of documented security objectives and controls from which to select as GAS goals. This does not necessarily compromise the methodology by reducing the effort of the goal setter because much care and attention is still necessary to select operational objectives, that is, IS controls, which are relevant and important to the organization. However, this effort will depend to a large extent on how relevant the set of IS security standards are to the organization.

Standards, in general, can either be generic or specific depending on the audience for which they are developed. IS security standards are no different. Examples of generic IS security standards are AS/NZS 17799 (Standards

Australia/Standards New Zealand, 2001), AS/NZS 4360 (Standards Australia/Standards New Zealand, 1999) and the Control Objectives for Information and related Technology (COBIT) (Information Systems Audit and Control Foundation (ISACF), 2000). These standards and guidelines were developed by teams of people from various industries with different knowledge bases for organizations in general. On the other hand, more specific standards have been developed in-house. For example, the standards used by the case study agency were developed for whole-of-government (South Australian Government, 1994b). Although this is not organization specific, since whole-of-government consists of approximately 140 agencies, at least they are South Australian government specific. It is reasonable to assume that the more specific an IS security standard is to an organization, the less chance there is of it containing irrelevant controls.

Irrespective of the degree of relevance of the set of IS security standards, the goal setter must decide which controls are important, which controls are objective-driven and how many controls will constitute a valid representation of the issue being evaluated. Some standards were found to be easier to work with than others. For example, the Australian/New Zealand Standard AS/NZS 17799 (Standards Australia/Standards New Zealand, 2001) provides a descriptive objective for each IS security category. Furthermore, objective-driven controls were not too difficult to identify simply by envisaging how their level of attainment could be scaled.

The problem of determining whether a selection of controls is a valid representation of the IS security category being evaluated can be in conflict with the desire to make the evaluation instrument a manageable size. The temptation is to include all objective-driven controls, thus eliminating the risk of omitting an important one. Kiresuk et al. (1994) suggest that the initial number of scaleable GAS goals for a single GAS follow-up guide is between 3 and 10.

More research needs to be done in this area of GAS goal setting for IS security because the "weakest link" phenomenon applies. For example, a suburban house may be considered to be adequately safeguarded against theft because it has multi-layered controls in the form of infrared sensors, window and door deadlocks and a guard dog. However, if valuables are not put in a safe, then they are most vulnerable to theft when visitors are in the house. The non-existence of this control represents the weakest link in the control set.

# Evaluating Processes or Outcomes

Kiresuk et al. (1994) point out that "GAS is appropriate only for outcome variables" (p. 31) and "In GAS, it is outcomes that are scaled" (p. 30). However, they accept that in certain circumstances the process or procedures undertaken to achieve the stated GAS goal may well be the operational objective. For example, consider the IS security control "Conduct an audit of physical access controls". If we were evaluating outcomes we could use as an indicator of measurement it would be the number of incidences of unauthorised physical access. On the other hand, if we were evaluating processes we might use as an indicator of measurement, it would be the frequency with which the audit is conducted. This equates to measuring the process put in place to achieve a goal rather than measuring an outcome of the process.

In order to implement GAS it is necessary to identify items to be evaluated as GAS goals. Are they to be outcomes, processes, enabling outcomes or some other form of output? One could be excused for thinking that the items to be evaluated in a goal-oriented evaluation of IS security controls would be control objectives such as "To reduce the number of viruses introduced" or "To minimise the number of unauthorised accesses to classified information," or "To reduce the number of accidental input errors". Objectives of this nature are certainly a valid item to assess. In contrast, this research focuses on evaluating the IS controls in place to achieve the objectives rather than evaluating the control objectives. This decision was made because the GAS methodology requires that the subject of evaluation (the GAS goal) be an objective that is scalable. The above-mentioned control objectives are certainly scalable but only in terms of the extent of security breaches that occur. In these cases the indicators of performance would need to relate to the number of breaches which actually occurred after the controls were put in place (or to be more accurate, the number of breaches which were prevented, avoided or detected). For example, one would have to estimate or record the number of viruses introduced or the number of unauthorised accesses that occurred or the number of accidental input errors that occurred. Alternatively, breaches of security could be measured by the impact on the business when they do occur. For example, the effect of a computer virus can range from "a bit of a laugh" to hundreds of thousands of dollars lost because the computer system was made inoperable for a significant period of time, causing hundreds of people to become unproductive.

Measurements of this nature may seem to be the most effective means of evaluating the extent to which the types of control objectives mentioned above have been attained; that is, to measure the outcome or impact of having implemented the IS control. However, there are problems associated with the measurement of security breaches. Firstly, accurate information pertaining to security breaches can be very difficult to collect for a variety of reasons. Management is often very reluctant to release information that may reflect on their management style. Security is a management responsibility and if there are too many incidents of security breaches, management may be viewed negatively. Guest (1962) referred to this problem when conducting studies of organization behaviour at the management level. He claims that managers are often reluctant to provide accurate information to external parties for fear of it being leaked to competitors or that it may upset a delicate power struggle occurring within. Further to this, it may not be prudent for details regarding security breaches to be made public in situations where the organization is, for example, a custodian of people's money or other personal valuables. Banks and financial institutions must maintain customer confidence or risk losing patronage.

Another problem with measuring security breaches to ascertain the state of an organization's information security is equally contentious. It relates to the number of occurrences of breach or the extent of damage caused by the breach not necessarily reflecting the quality of the controls in place. Security breaches can occur at any time. They do not necessarily occur more often when fewer controls are in place. Take, for example, natural disasters like fire, flood, earthquake or terrorist attack. These threats are an act of God and an organization with numerous high quality controls is just as vulnerable as an organization which has very few.

Consequently, the evaluation approach used herein has *not* attempted to measure security breaches as a means of assessing the state of IS security, but instead, has focused on the evaluation of the IS controls in place. This amounts to evaluating a process that will contribute to an outcome.

## The Need for an Independent Audit

It is important to appreciate that the GAS methodology espoused in this chapter is *not* an audit of the organization's IS controls. An IS audit would test inputs, processes, outputs and business operations for conformance against a standard, whereas the GAS methodology measures stakeholder perceptions of the current situation. Hence, it is recommended that the results be validated by conducting an independent audit of the controls.

The author is *not* questioning whether GAS can be used as a singular assessment method. However, it is generally accepted that alternate evaluation methods should be used to cross-validate (triangulate) data to increase the degree of inference about relationships (de Vries et al., 1992; Love, 1991; Patton, 1990, 1997; Steckler et al.,1992; all as cited in FIPPM, 1999). Another recent study (Robertson-Mjaanes, 1999) examined, among other things, the validity of GAS, and concluded that the best use of GAS may *not* be as a singular method. It was suggested that it is more suited when used to integrate and summarise the results obtained from other progress monitoring and outcome measuring methods as opposed to being used as a single assessment method.

Whether GAS is shown to be highly suitable or otherwise does *not* mean that it should be used on its own without an independent third party assessment to confirm results. In striving to increase the quality of outcome assessment (although in this research, the outcomes are in fact processes) it is important to complement the results of a GAS evaluation with another measure/assessment process. This is the advice given by Kiresuk et al. (1994), regardless of the domain in which it is being used. This is particularly relevant in the evaluation of any prevention program like a program to prevent human injury on the roads (see FIPPM, 1999), or indeed, a program of controls put in place to prevent breaches of IS security as in this research. This is because prevention programs of this type are seldom foolproof or 100% effective and furthermore, it is difficult to attribute a null result to causality of the intervention. The point is that there are an infinite number of controls that can be put in place and therefore choices have to be made as to which ones and how many. Consequently, it is important that more than one evaluation method is deployed to confirm that the best selection of controls and processes is made in order to maximise the effectiveness of the program.

It is important that some form of continuous monitoring occurs. One such approach is to adopt a global rating system whereby an overall assessment of the situation can be obtained. One such program is the Global Assessment Scale described by Hargreaves and Attkisson (Attkisson, Hargreaves, Horowitz & Sorensen, 1978), used in the evaluation of mental health programs. In the domain of IS security, continuous monitoring by an independent third party could be achieved by utilising the services of external auditors who currently have the responsibility of auditing IS controls. This would serve as a check on evaluator bias and generally help to improve the validity of the evaluation process.

As with the upstream petroleum environmental application (Malavazos & Sharp, 1997) the full third party assessment (e.g., consultant ecologist research over 12 months) can be highly time-consuming and expensive. In any case, prevention is more cost-effective, because during the course of the third party review the organization may have to endure high-risk operations to be properly tested. All of these considerations do not obviate the need for third party independent review and application of safe standards. However, they do heighten the need for regular, frequent, low cost self-evaluations to reduce the cost and risk of the fully-fledged review.

## The Need for GAS Training

The training of potential GAS goal setters is a critical component of the GAS methodology if a meaningful result is to be achieved. The GAS methodology is reported to have been used in hundreds of organizations and for many different purposes (Kiresuk et al., 1994). From these, Kiresuk et al. (1994) have concluded, "Effective training must be provided to staff who will be employing the technique, time must be allowed for the staff to use the technique properly, and administrative support must be available to sustain the implementation" (p. 6). In fact, proper training is considered such an important issue that Kiresuk et al. (1994) devoted a whole chapter to topics such as curriculum design, skills to be developed, GAS goal setting, GAS goal scoring and the costs of training.

## Evaluating a Non-Human Service

An important question is whether the same arguments for GAS's success in evaluating human service programs can be extended to a study where the program being evaluated is a management "program" (or plan) where the subjects are not individuals but procedures or activities. In the evaluation of a health or education program, the GAS methodology measures the outcomes of a program by assessing the impact that the program has on individuals; that is, the "condition" of the individuals. In contrast, this research uses a GAS-based methodology to measure the process of implementing and maintaining a set of

IS controls by assessing the state of these controls. The most significant difference between an evaluation of a human service compared to a non-human service appears to be that the recipient of human service may also be involved in the goal setting process. In the case of a non-human service, this does not happen because the recipient is the organization, not a human being. However, stakeholders or representatives of the organization should be involved in the goal setting process. Notwithstanding the limited number of non-human service test cases in which GAS has been applied, the author believes that the methodology warrants further investigation in these situations to ascertain its usefulness and relevance. At this stage it would be incorrect to assume that the success of GAS in the evaluation of human service delivery programs automatically applies to non-human service programs.

## Criticisms of GAS

Using the GAS methodology for the evaluation of IS security has its shortcomings and problems and is subject to criticism in the following areas:

- The approach does not offer solutions to poor IS security or recommend management action to improve IS security. This is the next step in the risk management process. This study is concerned with answering the management question "How secure are the ISs?" and does not attempt to address what should be done about them.

- The approach does not necessarily *audit* the security of ISs. It is a measure of stakeholder perceptions of the current situation. However, for more accurate results, it is recommended that an independent person audits the IS controls.

- Evaluation results are not necessarily foolproof and conclusions based on GAS summary scores need to be conservative. The domain of security and management controls is subject to a form of the weakest link phenomenon. In other words, an organization is only as secure as the weakest control. The GAS methodology generates an average score for each follow-up guide and since each follow-up guide consists of between three and ten controls, this means that strong controls could hide weak controls. For this reason it is recommended that all low scoring controls be followed up individually.

- Self-assessment approaches, by their very nature, often have a credibility problem in terms of evaluator bias and a lack of objectivity. This is particularly likely to occur in situations where (a) the evaluator is involved in the creation of the measuring device or (b) the evaluator is the person responsible for IS security within the organization. Irrespective of the conduct of the self-assessment approach, Love (1991) advocates that one of the key elements of any self-study process is that the achievement of standards or objectives should be confirmed by external evaluators (p. 81). For this reason it is recommended that independent persons, preferably external to the organization, audit the validity of the GAS instrument and the evaluation process.

- Sets of GAS scales are only individualised scales to detect change in the target organization and are *not* for generalisation across different organizations.

# Conclusion

This research showed that the GAS methodology, combined with a generally-accepted IS security standard, is not only a feasible approach for evaluating IS security but is also useful for assessing the level of compliance with that standard. This has particular relevance for public sector organizations that have a responsibility to comply with common whole-of-government standards.

In addition, this research highlighted a number of attributes of the GAS methodology that contribute to its suitability as a self-evaluation methodology when used to evaluate IS security. The most relevant of these attributes is the fact that the GAS evaluation instrument is developed by stakeholders within the organization and used periodically by other stakeholders to monitor the level of IS security against an established standard.

# Recommendations

This research has spawned the following key recommendations relating to the use of the GAS methodology when used to assess IS security:

- A set of formally documented IS security standards should be used as a basis for developing the GAS evaluation instrument. This will:
  - assist in identifying the IS security categories to be evaluated;
  - ensure that the IS controls selected are truly representative of all the controls within a particular IS security category; and
  - assist goal setters in developing an appropriate scale of GAS goals that are measurable.

- A generic or core instrument for whole-of-government should be developed before individual agencies develop their own individualised instruments. This core instrument, based on whole-of-government IS security standards, should then be used as a starting point for individual agencies.

- IS security should be assessed against a baseline set of controls such as a policy document or a set of standards. This implies that GAS should be used to measure the extent of compliance with a baseline set of controls, rather than attempting to assess the adequacy of existing controls.

- The selection of IS controls and the subsequent goal setting process is not only the most difficult aspect of the GAS methodology, but it is also the most important. It is therefore recommended that an independent person audit the IS controls and the five descriptive levels of attainment of each control.

- The selection of controls to be included in each GAS follow-up guide is a critical process that can determine the validity of the evaluation results. In order to select a truly representative set of controls for the IS security category being assessed, it is recommended that objective-driven controls are chosen in preference to compliance-driven controls.

## Chapter Summary

This chapter has described how an established and well-accepted methodology in one discipline can be adapted to a totally different discipline and still maintain its merit. Although this research is limited and quite specific as a single case study, it has extended the research of Malavazos and Sharp (1997), who used the methodology to assess the environmental performance of the petroleum industry. This research also extended the research of von Solms, van de

Haar, von Solms and Caelli (1994), who claimed that "...no information security self-evaluation tool, known to the authors, exists currently" (p. 149). At that time, they were not aware of any internationally accepted IS security standards. Since then, a number of such standards have been published (e.g., AS/NZS 17799, BS7799). As a consequence, it seems that it is now time for IS security research to concentrate on validating existing techniques and developing new techniques.

This research needs to be extended to enable the following questions to be answered:

- Is it suitable for different sized organizations?
- Is it suitable in organizations with multiple IT platforms?
- How can staff be trained to overcome the need for a GAS methodologist?
- Is the effort in developing a GAS measuring instrument a better utilisation of time and resources than actually taking preventative action?

# References

Attkisson, C.C., Hargreaves, W.A., Horowitz, M.J., & Sorensen, J.E. (Eds.). (1978). *Evaluation of human service programs.* New York: Academic Press.

FIPPM. (1999). *Evaluating injury prevention initiatives.* Developed by Flinders Institute of Public Policy and Management (FIPPM), Research Centre for Injury Studies, Flinders University, South Australia, 1-8.

Guest, R.H. (1962). *Organizational change: The effect of successful leadership.* London: Tavistock Publications.

Information Systems Audit and Control Foundation (ISACF). (2000). *Control Objectives for Information and related Technology (COBIT)* (3rd ed.). USA.

Isaac, S., & Michael, W.B. (1995). *Handbook in research and evaluation* (3rd ed.). San Diego, CA: Edits Publishers.

Kiresuk, T.J., & Lund, S.H. (1982). Goal attainment scaling: A medical-correctional application. *Medicine and Law, 1,* 227-251.

Kiresuk, T.J., Smith, A., & Cardillo, J.E. (Eds.). (1994). *Goal attainment scaling: Applications, theory and measurement.* NJ, USA: Erlbaum Inc.

Love, A.J. (1991). *Internal evaluation: Building organizations from within.* Newbury Park, USA: Sage Publications.

Malavazos, M., & Sharp, C.A. (1997, October). Goal attainment scaling: Environmental impact evaluation in the Upstream Petroleum Industry. *Proceedings of Australasian Evaluation Society 1997 International Conference,* Adelaide, South Australia, 333-340.

Owen, J.M. (1993). *Program evaluation: Forms and approaches.* NSW, Australia: Allen & Unwin.

Robertson-Mjaanes, S.L. (1999). An evaluation of goal s as an intervention monitoring and outcome evaluation technique. Dissertation at University of Wisconsin-Madison, published by UMI Dissertation Services.

South Australian Government. (1994a, July). *South Australian government information technology security guidelines* (vol. 1).

South Australian Government. (1994b, December). *South Australian government information technology security standards in an outsourced environment* (vol. 1).

Standards Australia/Standards New Zealand. (1999). *Risk management,* AS/NZS 4360:1999. Strathfield, NSW, Australia: Standards Association of Australia.

Standards Australia/Standards New Zealand. (2001). *Information technology - Code of practice for information security management,* AS/NZS ISO/IEC 17799:2001.

von Solms, R., van de Haar, H., von Solms, S.H., & Caelli, W.J. (1994). A framework for information security evaluation. *Information and Management, 26*(3), 143-153.

**Chapter XII**

# Information Security Policies in Large Organisations:
## The Development of a Conceptual Framework to Explore Their Impact

Neil F. Doherty
Loughborough University, UK

Heather Fulford
Loughborough University, UK

## Abstract

*While the importance of the information security policy (ISP) is widely acknowledged in the academic literature, there has, to date, been little empirical analysis of its impact. To help fill this gap a study was initiated that sought to explore the relationship between the uptake, scope and dissemination of information security policies and the accompanying levels of security breaches. To this end, a questionnaire was designed, validated and then targeted at IT managers within large organisations in the United Kingdom. The aim of this chapter is to provide a progress report on this study by describing the objectives of the research and the design of the conceptual framework.*

"*I only ask for information.*" (From David Copperfield (Chapter 20) by Charles Dickens)

# Introduction

For the past two decades it has been argued that an "*information revolution*" is taking place that is having a significant impact upon all aspects of organisational life (e.g., Drucker, 1988; Porter & Millar, 1985). If applied effectively as a strategic resource, information investments can result in the realisation of significant corporate benefits; indeed it has been contended that "information is the lifeblood of the organisation" (CBI, 1992, p.2). It can be argued that information is vital to the success of the business, as it contributes directly to the organisation's operational performance and financial health (Bowonder & Miyake, 1992; McPherson, 1996). However, information will only be recognised as a vital organisational resource if managers can readily gain access to the information they require. Like the character from David Copperfield, in the introductory quotation, many managers are desperate to gain access to the information they need. Unfortunately, as a consequence of the high incidence of security breaches, many organisations are failing to consistently provide the information resources that their managers require (Angell, 1996; Gaston, 1996).

Organisations must make every effort to ensure that their information resources retain their integrity, confidentiality and availability. However, the increasing integration of information systems both within and between organisations, when coupled with the growing value of corporate information resources, have made information security management a complex and challenging undertaking (Gerber et al., 2001). Indeed, it is estimated that "security breaches affect 90% of all businesses every year, and cost some $17 billion" (Austin & Darby, 2003, p.121). Moreover, Austin and Darby (2003) also suggest that protective measures can be very expensive: "the average company can easily spend 5% to 10% of its IT budget on security". One increasingly important mechanism for protecting corporate information, in an attempt to prevent security breaches, rather than respond to them, is through the formulation and application of an information security policy (ISP) (Hone & Eloff, 2002).

While the high incidence of security breaches and the importance of information security policies are both areas that have attracted significant attention in the literature, there is little evidence that these topics have been explicitly combined. To help fill this gap a research study was initiated that sought to empirically explore the relationship between the uptake and application of information security policies and the incidence of security breaches. The aim of this chapter is to provide a progress report on this study by providing a thorough review of the literature, describing the objectives of the research and presenting the conceptual model. The remainder of this chapter is organised into four sections: a review of the literature, a discussion of the research objectives and the conceptual framework, a description of the methods used, and finally, the conclusions and recommendations for future research.

# Literature Review

The aim of this section is to present a discussion of the literature with regard to the value of information, the threats to the security of information, which can greatly diminish its value, and the role of the information security policy in countering such threats. The section concludes with a critique of this literature, and in so doing, establishes the academic justification for this research.

## The Value of Information

Over the past 20 years or more an important body of literature has emerged that examines the many ways in which information has the potential to contribute to organisational performance. For example, McPherson (1996, p.203) argues that "information is vital to the success of the business and will be accountable for a significant share of the business's various indicators of success, including its cash flow and market value". Some of the more common themes from this body of literature (e.g., Bowonder & Miyake, 1992; Davenport & Cronin, 1988; Porter & Millar, 1985) include:

- Information facilitates effective decision-making;
- information supports the day to day operations of the organisation;
- information is a product in its own right;
- information is the critical ingredient in strategic planning;
- information is incorporated into many products and services to enhance their usability and value.

The value-adding potential of information has increased to the extent that Glazer (1993, p. 99) argues that information should now be viewed as the "firm's primary strategic asset". However, Framel (1993) stresses that information is "an asset that needs to be managed in order to maximise value and return to the organisation". Unfortunately, in practice, too many organisations view information as an overhead that has to be managed as a cost rather than as an asset (Orna, 1999; Strassmann, 1985). Consequently, many organisations are reluctant to make an appropriate investment in information resources, and information exploitation activities are often the first to be cut back when business is in decline (Orna, 1999). This failure to treat information as a valuable asset and manage it accordingly means that too often "invaluable corporate information assets lie like sediment at the bottom of an ocean" (Rostick, 1994, p.24).

One of the most likely reasons for information being under-utilised, and not therefore being perceived as valuable, as is discussed in the following section, is that far too often its availability, integrity or confidentiality are compromised by security breaches (BSI, 1999; Menzies, 1993). Consequently, if the security of information and the systems on which it is processed and stored can be improved, then there is the potential for substantially enhancing the value of information being generated and utilised within the organisation.

## The Importance of Information Security

Given the growing importance of information, it is not perhaps surprising that the security of information is becoming an increasingly important organisational concern. Indeed, a growing, and increasingly influential body of literature on the security of information has been generated over the past 20 years. It should be noted that this body of literature includes contributions that have been labeled "computer security" or "IT security," as well as "information security". However, the term "information security" is preferred, and will be used throughout the remainder of this chapter, because it is the information, rather than the technological infrastructure, which as the "key organisational resource" (Gerber, von Solms & Overbeek, 2001, p. 32), has the potential to deliver value.

Information only retains the potential to deliver value, if its confidentiality, integrity and availability can be protected (Gaston, 1996, p. 2; Menzies, 1993, p. 164). Recent surveys of information security issues suggest that awareness has increased among senior managers of the value of information and the need to protect it (see Hinde, 2002 for a summary of these surveys). Unfortunately, there is a significant, and growing, range of factors that threaten the security of organisational information. Of these threats, the following eight are probably the most significant, and therefore worthy of further discussion:

- *Theft:* As the value of both the hardware and software of computer systems increases, so too does their attractiveness to thieves. In particular, the theft of computer chips is becoming common, as they are relatively small, yet highly valuable (Bocij et al., 1999, p. 538). In addition, theft of data is now recognised to be particularly damaging to an organisation's effectiveness.

- *Computer-based fraud:* The most common type of fraud is where data processing or data entry routines are modified so that there is scope to embezzle the target organisation. There is growing evidence that computer-based fraud is widespread: over 90% of companies responding to a survey had been affected (Romney, 1996). Moreover, the economic losses from such cases can be "staggering," with average losses estimated to be in excess of $100,000 (Haugin & Selin, 1999, p.340).

- *Computer virus:* A recent worrying external threat facing organisational information and systems has been the rapid proliferation of viruses, worms and trojans (Post & Kagan, 2000). These increasingly common threats are computer programs that have the capability to automatically replicate themselves across systems and networks, as well as typically delivering malevolent or mischievous functionality.

- *Hacking incidents:* Another significant external threat is the penetration of organisational computer systems by hackers. Such attacks, often termed "intrusions" (Austin & Darby, 2003, p.122), can be particularly dangerous, as once the hacker has successfully bypassed the network security, he/she is free to damage, manipulate or simply steal data at will. Related to this aspect of security threat is cyber-terrorism, incorporating, for example, unlawful attacks designed to intimidate or coerce (Hinde 2003).

- *Human error:* Where systems are poorly designed and maintained, users are inadequately trained and/or control procedures are lax, there is always a significant threat of security breaches as the result of human error. Indeed, human error, whereby computer users either accidentally enter incorrect data, or accidentally destroy existing data, was perceived to be by far the most significant security problem facing U.S. organisations (Loch et al, 1992). More recently, a DTI survey of information security found human error to be a considerable threat to the availability, confidentiality and integrity of information (DTI, 2000).

- *Damage by disgruntled employees:* One potentially very damaging employee response to being dismissed, overlooked for promotion or simply aggravated by a manager is to seek revenge by damaging their employer's computer systems or the data contained within them. The scale of the problem was highlighted in a recent survey, which found that a third of all responding companies felt that their corporate information security was at risk from disgruntled employees (Mitchell et al., 1999).

- *Unauthorised access to/use of data (internal):* It has been noted in recent surveys that there has been a change in security incidents in organisations in recent years, such that external incidents now seem to outweigh the earlier more predominant internal threats (Hinde, 2002). Nevertheless, internal threats still exist, and can be defined as "the intentional misuse of computer systems by users who are authorized to access those systems and networks" (Schultz, 2002, p.526). Such internal incidents can be particularly damaging if they result in the misuse of confidential or commercially sensitive data.

- *Natural disaster:* Whilst the vast majority of threats to information and information technology originate from human perpetrators, they are also susceptible to non-human threats, most typically in the form of natural disasters (Loch et al., 1992). Natural disasters come in many guises: their most common manifestations are phenomena such as floods, earthquakes or fires, which destroy computing facilities, or lightning strikes and power surges that destroy data.

Unfortunately, the threats to information resources continue to grow, primarily as a result of the high levels of interconnectivity witnessed both within and between organisations (Barnard & von Solms, 1998; Dinnie, 1999; DTI, 2002). In particular, the increased incidence of inter-organisational systems is creating many new problems for organisations. Information security is no longer solely a "domestic" issue, but one which now also involves external business partners (von Solms, 1998, p. 174). The high levels of connectivity offered by the Internet have raised many new concerns about information security: Internet users are not only potential customers or suppliers; they are also potential security threats (Pfleeger, 1997). Indeed, it has been reported that the fear of security breaches constitutes the greatest inhibitor to the uptake of electronic commerce (Ernst & Young, 2001). Increased interconnectivity is not, however, the only factor heightening the importance of security; the growing organisational importance of information value, as discussed earlier, has also brought information security nearer to the top of the management agenda (Gerber et al., 2001).

When the availability, integrity or confidentiality of an information system is compromised by one of the breaches discussed earlier, the costs in terms of direct losses tend to be modest, typically only 5% of the aggregate cost of security breaches (Menzies, 1993). By contrast, the consequential losses of security breaches can be huge, accounting for approximately 95% of the total costs. As Austin and Darby (2003, p.121) note: "security breaches can have far reaching business implications, as they can disrupt operations, alienate customers and tarnish reputations". Information security is not, therefore, simply a technical issue that is of concern only to IT practitioners; it should be the concern of all employees, particularly managers. One mechanism for placing information security on the agenda of all employees is through the formulation and application of an information security policy.

## The Role of the Information Security Policy

The information security policy (ISP) has become the focus for an increasing amount of academic scrutiny, particularly over the past ten years. The aim of this section is to provide a broad review of this body of literature, paying particular attention to the importance, types and uptake of policies. However, we start this section by exploring the definition of an ISP. Gaston (1996, p. 175) suggests that the ISP can be defined as "broad guiding statements of goals to be achieved; significantly, they define and assign the responsibilities that various departments and individuals have in achieving policy goals". This definition is broadly in line with the British Standard on Information Security Management (BSI, 1999, p.3), which suggests that the ISP document should "set out the organisation's approach to managing information security". As such, ISPs typically include "general statements of goals, objectives, beliefs, ethics and responsibilities, often accompanied by the general means of achieving these things (such as procedures)" (Wood, 1995, p.668). Wood (1995, p.668) also makes the important distinction between security policies and guidelines: "policies are mandatory," and consequently, "special approval is needed where a worker wishes to take a different course of action," whilst guidelines tend to be advisory.

Whilst there is a high degree of consensus within the literature with regard to a broad definition for the ISP, there is rather more debate as to whether there should be a single policy, or whether it should be subdivided into several distinct levels or types (Baskerville & Siponen, 2002). For example, Siponen (2000a, p.111) suggests that security policies can be classified into two broad groups, namely "computer-oriented policies" or "people/organisational policies". Sterne (1991, p.337) distinguishes between three levels of policy, namely the "institutional policy, the institutional ISP and the technical ISP". While Lindup (1995, p.691) agrees that there is no single ISP, he suggests there are several distinct types, rather than levels, which include: "system security policy, product security policy, community security policy and corporate information security policy". Academics may continue the debate about ways of classifying or sub-dividing policies, but in practice organisations tend to have a single "corporate policy" (Lindup, 1995, p.691). Consequently, for the purposes of the study described in this chapter, it is assumed that organisations are only likely to have a single information security policy. However, we recognise that in practice such policies are likely to be supported by a variety of lower level "standards" (Wood, 1995, p.668) or "procedures" (Dhillon, 1997, p.3), which tend to have a more technical or operational perspective. Policies, therefore define "what is required" (Moule & Giavara, 1995, p.16), whilst the supporting standard or procedure typically describes "how to do it".

There is a growing consensus within both the academic and practitioner communities that the information security policy (ISP) is the basis for the dissemination and enforcement of sound security practices within the organisational context (Baskerville & Siponen, 2002). As David (2002, p.506) notes: "it is well known, at least among true security professionals, that formal policy is a prerequisite of security". Similarly, Lindup (1995, p.691) asserts:

*ten years ago, information security policies were more or less unheard of outside the world of secret military and government networks. Now they are regarded by security professionals as one of the most important foundations of information security.*

The primary reason that the ISP has become the "prerequisite" (David, 2002, p.506) or "foundation" (Lindup, 1995, p.691) of effective security practices has been suggested by Higgins (1999, p.217), who notes: "without a policy, security practices will be developed without clear demarcation of objectives and responsibilities".

However, there is also a growing concern that too many organisations are failing to heed this advice, as witnessed by the low levels of uptake of formal information security policies (Arnott, 2002), and the inadequacies in policies where they do exist (Hone & Eloff, 2002; Moule & Giavara, 1995). Whilst the importance of, and concerns about, information security policy are widely recognised, this interest has not, as yet, been translated into detailed empirical surveys explicitly targeting the utilisation of information security policies in organisations.

Another important strand of the literature concerns the uptake of ISPs. There have been a number of recent surveys addressing this issue, although it is interesting to note that these have all been practitioner, rather than academic, studies. Whilst these studies addressed a broad range of security issues, they focused, in particular, on the prevalence of ISPs within European organisations. For example, the Andersen study (2001) reports that 65% of the organisations surveyed (most of which were large organisations) had an information security policy in place. Similarly, although a UK-based survey (DTI, 2002) reported that only 27% of its sample had a policy in place, it also noted that 59% of large organisations had implemented a policy. The DTI (2002) survey also noted a strong upward trend in the adoption of ISPs; an earlier study (DTI, 2000) reported that only 14% of the organisations surveyed had an ISP in place. Moreover, the 2002 study found that a higher proportion of organisations with a policy were undertaking annual policy updates than had been the case in 2000.

## Critique of Literature

This review of the literature has found that the importance of information security, and in particular the ISP, is increasingly being recognised. However, a number of significant gaps still exist within the literature. For example, little can be found within the literature about the scope, updating or longevity of information security policies. Moreover, much of our current understanding of the role of the ISP comes from the practitioner literature, which whilst it makes an important contribution, it is not typically subjected to the same level of scrutiny as formal academic studies. Dhillon and Backhouse (2001) provide a strong indication of the relative paucity of academic literature focussing upon the ISP. Their comprehensive review of the information security literature concluded that existing research tends to focus upon "checklists [of security controls to be implemented], risk analysis and evaluation"; information security policy was not explicitly featured in their review. This is an important gap in the literature, because as David (2002) acknowledges, the formulation of an ISP is the critical element that should explicitly link the assessment of risk and the implementation of security controls. There is, therefore, a pressing need for more academic exploration of this area, particularly empirical studies that focus explicitly on the uptake, role and application of the ISP. By contrast, one aspect of information security that has been well investigated is the nature and incidence of security breaches. However, to date, there has been little conceptual or empirical scrutiny to determine whether the incidence of security breaches can be reduced through the adoption of an information security policy.

# Research Objectives and the Conceptual Framework

The aim of this section is to describe the study's broad objective before articulating the specific research hypotheses and presenting the research framework. It should be noted that a full discussion of the design of the questionnaire and the operationalisation of the constructs discussed in this section is deferred to the following section. Given the lack of empirical research in the area it was felt that an exploratory piece of work that embraced a wide range of issues would be most appropriate. To this end the aim of the study was to explore how a variety of issues relating to the uptake and application of information security policies impacted upon the incidence of security breaches within large organisations. Based upon our review of the literature, it is possible to hypothesise that a number of distinct aspects of the ISP might influence the incidence of security breaches. Each of these areas is represented as a significant construct on the conceptual framework (see Diagram 1), and each can be linked to a research hypothesis, as described below:

- *The existence of an ISP:* The literature is very clear that the ISP is an important prerequisite of effective security management (Baskerville & Siponen, 2002; David, 2002; Lindup, 1995). The corollary of this is that the formulation and application of a formal, documented ISP should lead to a reduction in security breaches. The following hypothesis is therefore proposed:

    **H1:** Those organizations that *have a documented ISP* are likely to have fewer security breaches, in terms of both frequency and severity, than those organisations that *do not.*

- *The age of the ISP:* Whilst the literature has relatively little to say about the importance of the longevity of IS policies, it may be assumed that organisations with a long history of utilising such policies might be more effective in the management of information security. In particular, it might be hypothesised, as follows, that the more experienced users of ISPs might experience fewer security breaches:

    **H2:** Those organizations that have had an ISP in place for *many years* are likely to have fewer security breaches, in terms of both frequency and severity, than those organisations that *have not.*

- *The updating of the ISP:* The literature also has relatively little to say about how frequently the ISP should be updated. However, it may be assumed that in those organisations where security is high on the agenda the ISP might be updated more frequently, which in turn might improve the effectiveness of an organisation's security management. More specifically, it might be hypothesised, as follows, that the frequent updating of an ISP might result in fewer breaches.

    **H3:** Those organizations that update their ISP *frequently* are likely to have fewer security breaches, in terms of both frequency and severity, than those organisations that *do not.*

- *The dissemination of the ISP:* There is a strong consensus within the literature that policies will only be effective if they are communicated throughout the organisation (BSI, 1999; David, 2002; Siponen, 2000b). The natural implication of this is that the dissemination of the ISP might impact upon the frequency and severity of security breaches. It is, therefore, possible to propose the following hypothesis:

  **H4:** Those organizations that *actively disseminate* their policy are likely to have fewer security breaches, in terms of both frequency and severity, than those organisations that *do not.*

- *The Scope of the ISP:* Whilst the literature has relatively little to say about the specific issues to be covered by the ISP, it seems reasonable to assume that there may be a relationship between the scope of the ISP and the effectiveness of an organisation's security management. In particular it might be hypothesised, as follows, that a broad scope might be associated with fewer breaches:

  **H5:** Those organizations that have a policy with a *broad scope* are likely to have fewer security breaches, in terms of both frequency and severity, than those organisations that *do not.*

- *The adoption of success factors:* The British Standard (BSI, 1999) has some very clear advice about the factors that are important in ensuring the successful application of an ISP. The importance of many of these factors is under-pinned by their discussion in the wider academic literature (e.g., Moule & Giavara, 1995; Wood, 1995). The corollary of this is that the adoption of these success factors should lead to a reduction in security breaches. The following hypothesis can therefore be proposed:

  **H6:** Those organizations that have adopted a wide variety of *success factors* are likely to have fewer security breaches, in terms of both frequency and severity, than those organisations that *have not.*

While the hypotheses have been formulated to represent the outcomes that the researchers believed to be the most likely, it was recognised that in some cases alternative, yet equally plausible, results might be produced. For example, it might be that the existence of an ISP is associated with a high incidence of security breaches in circumstances in which the policy has been implemented in direct response to a poor security record.

*Figure 1. Conceptual model of study*

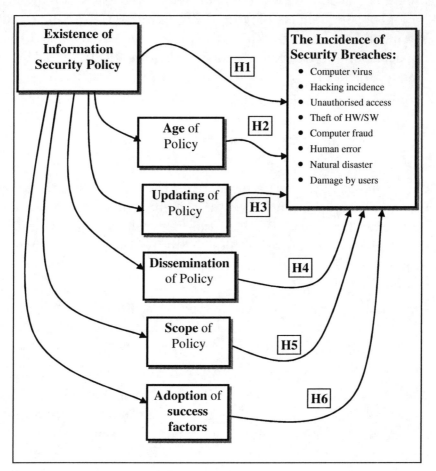

## Research Method

To effectively explore the research hypotheses, it was necessary to develop a series of measures that, when incorporated into a questionnaire, would adequately describe an organisation's information security activity. To this end, the questionnaire was designed through an iterative process of review and refinement. It sought to capture a significant amount of information with regard to the respondent's organisation, in addition to the information required to explicitly address the six research hypotheses.

# Questionnaire Development and Validation

A draft questionnaire was developed, based primarily upon the results of the literature review summarised in Section 2. As there are few published academic papers explicitly addressing the use of information security policies, the literature was used primarily to generate ideas and insights, rather than as a source of specific questions and item measures that could be utilised directly in this study. The resultant questionnaire was organised into the following five sections:

1. **Demographics:** The initial questions sought to build a demographic profile of the responding organisation. Information relating to organizational size, geographical spread and sector was collected so that the potential moderating effect on the statistical analyses could be explored.

2. **The incidence of security breaches:** This was operationalised as a multi-dimensional construct. A number of potential risks to the security and integrity of computer-based information systems were identified from the literature, as discussed in section 2.2, and included in the survey. A total of eight distinct threats, including computer viruses, hacking, human error, fraud and natural disasters, were identified and ultimately included in the research instrument. Each of these threats was operationalised in the following two ways:

   1) **Frequency of breaches:** Respondents were asked to estimate the approximate number of occurrences of a specified threat that they had experienced in the previous two years, using a four-item ordinal scale [**0**; **1-5**; **6-10**; **>10**].

   2) **Severity of breaches:** Respondents were also asked to estimate the severity of the worst incident, over the same two-year period, using a five-point Likert scale [**1** = fairly insignificant; **5** = highly significant].

3. **The existence and dissemination of the information security policy:** This section sought to determine whether a responding organisation had a documented information security policy, and if it did, how long the policy had been in existence, how often it was updated, and how the policy was disseminated. More specifically, the following topics were addressed:

   • **The existence of a policy:** The point of departure for this study was to ascertain whether a participating organisation had formulated a documented ISP. Consequently, on the questionnaire the existence of an ISP was operationalised as a simple dichotomous variable. If the organisation did have a policy the remaining questions were also then asked.

- **The age of the policy:** If an ISP was in use, respondents were asked to specify the number of years that it had actively been in operation.

- **The updating of the policy:** Respondents were also asked to identify the frequency with which the policy was typically updated, using a five-point, ordinal scale [> every two years; every two years; every year; every six months; < every six months].

- **The dissemination of the policy:** Policies are of little use unless all employees are made aware of their rights and responsibilities in relation to it. In addition to explicitly asking whether policies were disseminated via a *company intranet* or the *staff handbook,* respondents were asked to stipulate any *other* dissemination mechanisms being employed by the organisation.

4. **The scope of the policy:** Policies may vary greatly in their scope. A great deal has been written about the types and levels of security policy (see section 2), but there is relatively little academic material that explicitly addresses the scope or content of security policies. Ultimately, a very useful list of issues to be addressed through a policy was found on the Web. This was published electronically, as a white paper, by the UK-based organisation Claritas Information Security (*http:// www.centurycom.co.uk*), and it was used as a point of departure for creating a typology of policy scope items. As can be seen from the list of scope items presented in Table 1, many of these items have also been highlighted in the paper-based literature, although not in the form of an explicit list. For each of these issues, the respondent was invited to indicate whether the issue was covered in "the policy document *only,*" "a stand-alone procedure *only,*" "the policy document *and* a supplementary procedure," or if the issue is "not explicitly covered". It is anticipated that by summating the number of issues explicitly covered by the policy it will be possible to categorise the scope of the policy as either: broad (8-11 issues), intermediate (4-7 issues) or narrow (0-3 issues).

5. **The adoption of success factors:** It has been suggested that organisations will only be successful in the adoption of their ISP if they apply a range of

*Table 1. The scope of the information security policies*

| IT security issue | Reference Sources |
|---|---|
| Personal usage of information systems | BSI, 1999 |
| Disclosure of information | |
| Physical security of infrastructure and information resources | BSI, 1999 |
| Violations and breaches of security | |
| Prevention of viruses and worms | BSI, 1999 |
| User access management | BSI, 1999 |
| Mobile computing | |
| Internet access | Higgins, 1999 |
| Software development and maintenance | BSI, 1999 |
| Encryption | Wood, 1996 |
| Contingency/continuity planning | BSI, 1999 |

success factors (BSI, 1999). The British Standard identifies eight distinct factors, as highlighted in Table 2. As Wood (1996) notes: "no matter how sophisticated the information security technology, controls will not be sustainable unless the human element has been equally addressed". It is interesting that an inspection of the factors in Table 2 suggests that the success of the policy is predominately predicated on a range of human and organisational, rather than technical, issues. For each of these factors, the respondent was asked to indicate its importance, and the extent to which his/her organisation was successful in adopting that factor, using two separate five-point Likert scales.

The draft questionnaire was initially validated through a series of pre-tests, first with four experienced IS researchers, and then after some modifications it was re-tested with five senior IT professionals, all of whom had some responsibility for information security. The pre-testers were asked to critically appraise the questionnaire, focusing primarily on issues of instrument content, clarity, question wording and validity, before providing detailed feedback via interviews. The pre-tests were very useful, as they resulted in a number of enhancements being made to the structure of the survey and the wording of specific questions. Having refined the questionnaire, a pilot study was also

*Table 2. Success factors for the adoption of information security policies*

| Factors | Reference Sources |
|---|---|
| Visible commitment from management | BSI, 1999; Wood, 1995; Moule & Giavara, 1995 |
| A good understanding of security risks | BSI, 1999 |
| Distribution of guidance on IT security policy to all employees | BSI, 1999 |
| A good understanding of security requirements | BSI, 1999 |
| Effective marketing of security to all employees | BSI, 1999; Wood, 1995 |
| Providing appropriate employee training and education | BSI, 1999 |
| Ensuring security policy reflects business objectives | BSI, 1999 |
| An approach to implementing security that is consistent with the organizational culture | BSI, 1999; Moule & Giavara, 1995 |
| Comprehensive measurement system for evaluating performance in security management | BSI, 1999 |
| Provision of feedback system for suggesting policy improvements | BSI, 1999 |

undertaken, which provided valuable insights into the likely response rate and analytical implications for the full survey.

## Questionnaire  Targeting

It was recognised that only those individuals who had a high degree of managerial responsibility for information systems and technology would be able to comment knowledgeably about the uptake and scope of information security policies. Senior IT managers were, therefore, chosen as the "key informant," as they would be able to provide the requisite perspective, as these were judged to be most likely to be responsible for the formulation and application of an ISP. Moreover, only large organisations (firms employing more than 250 people) were targeted, as previous research has found that small firms tend to have few, if any, dedicated IT staff (Prembukar & King, 1992). To this end, a list of the addresses of IT directors from large UK-based organizations was purchased from a commercial market research organization. Each of the sample of 2838 IT directors was mailed a questionnaire, with an accompanying letter that explained the study, and a pre-paid envelope.

## Planned Analysis Strategy

As portrayed in the conceptual framework (Figure 1) and articulated in the research hypotheses, it is anticipated that there may be important relationships between each of the six independent variables relating to the uptake and application of the ISP, and the dependent variable: incidence of security breaches. To explore these relationships, it is envisaged that the data will be analysed using either ANOVA or Pearson correlation, depending upon whether the independent variables have been operationalised as ordinal or metric scales. However, it is anticipated that where a variable has been operationalised as a multi-item scale, as in the cases of "scope of policy," "success factors" and "incidence of breaches," it may be necessary to reduce the data, through the application of factor analysis or cluster analysis, before the hypotheses are tested.

A further important strand of analysis relates to exploring whether any of the findings, or indeed relationships, are contingent upon the organisational characteristics, as measured by the demographic variables. As Baskerville and Siponen (2002, p.338) acknowledge, simply following the prescriptions of general standards such as BS 7799 (BSI, 1999) can be dangerous, as they "do not pay adequate attention to the fact that all organisations are different". It might, for example, be that the adoption of an ISP has a more important role to play in reducing security breaches in certain sizes or types of organisations.

## Progress Report

At this point in time, the full survey has now been distributed, targeting 2838 senior IT executives in total. Of the questionnaires mailed out, only 208 valid responses have thus far been received, representing a response rate of 7.3%. This initial response rate is somewhat disappointing, as it does not allow us to confidently explore all of the hypotheses. In particular, the application of cluster and factor analysis to the multi-item scales could be undertaken with more confidence if there were more responses (Hair et al., 1998). A follow-up mailing is underway in an attempt to generate more responses. It is envisaged that the statistical analysis of the research hypotheses will begin in the near future and should generate some very interesting results. In the meantime, we have already published a provisional analysis of the results of the survey data (Fulford & Doherty, 2003) that presents a descriptive analysis of the data relating to the application of ISPs, but it does not describe or test any of these hypotheses.

While the initial response rate is rather disappointing, it is not perhaps surprising given the increasingly sensitive nature of information security. As Menzies (1993) notes, "organisations are reluctant to publish details of security incidents which have occurred, partly because of the knock-on effect this might have on shareholder, customer and public confidence, and also because of the embarrassment caused by making such events public knowledge."

# Concluding Remarks

There is a high degree of consensus, within the literature, that effective information security management is predicated upon the formulation and utilisation of an information security policy. However, there has to date been little empirical work to test this hypothesis. This chapter presents a research framework, with an accompanying set of hypotheses, that is designed to explicitly test whether, and under what circumstances, the adoption of an information security policy is likely to reduce the incidence of security breaches within large organisations. Moreover, the study should also provide very useful empirical insights with regard to the uptake, scope and dissemination of such policies. From a practitioner viewpoint, the results of this study should be very important, as it is anticipated that the findings will help organisations to better understand the value of security policies and to pinpoint the policy areas for prioritisation. The study also has important implications for the researcher, as it presents a set of constructs and item measures that can be adapted for use in other information security studies. Moreover, it is envisaged that the results of the study will play an important role in bridging and unifying the existing bodies of literature relating to security threats and the role of the ISP.

Social inquiry within the organizational context is always an ambitious undertaking, and therefore contains a number of inherent limitations. In particular, the adoption of the survey format restricts the range of issues and constructs that can be explored. There is also the potential for response bias associated with targeting only managerial stakeholders. It must also be recognised that as the results of this study will be based upon statistical analysis, they will be measuring *association* rather than *causality*. As a result, it is recognised that the quantitative study will not provide all the answers with regard to the impact of

ISPs. A series of follow-up interviews is, therefore, planned to provide deeper insights into the nature of any significant relationships that the quantitative analysis might uncover.

# References

Andersen (2001). *Sicherheit in Europa, Studie 2001, Status Quo, Trends, Perspektiven*, Andersen, Dussledorf.

Angell, I.O. (1996). Economic crime: Beyond good and evil. *Journal of Financial Regulation & Compliance, 4*(1).

Arnott, S. (2002). Strategy Paper. *Computing*, 28th February, 16.

Austin, R.D., & Darby, C.A. (2003). The myth of secure computing. *Harvard Business Review*, June.

Barnard, L., & von Solms, R. (1998). The evaluation and certification of information security against BS 7799. *Information Management and Computer Security, 6* (2), 72-77.

Baskerville, R., & Siponen, M. (2002). An information security meta-policy for emergent organisations. *Information Management and Computer Security, 15*(5/6), 337-346.

Bocij, P., Chaffey, D., Greasley, A., & Hickie, S. (1999) *Business Information Systems*, London: Financial Times Management.

Bowonder, B., & Miyake, T. (1992). Creating and sustaining competitiveness: Information management strategies of Nippon Steel Corporation. *International Journal of Information Management, 1*(3), 155-172.

B.S.I. (1999). *Information security management -BS 7799-1:1999*, London, UK: British Standards Institute.

C.B.I. (1992). *IT The Catalyst for Change*, London: Confederation of British Industry.

Davenport, L., & Cronin, B. (1988). Strategic information management - Forging the value chain. *International Journal of Information Management, 8*(1), 25-34.

David, J. (2002). Policy enforcement in the workplace. *Computers and Security, 21*(6), 506-513.

Dhillon, G. (1997). *Managing information systems security*. London: Macmillan Press.

Dhillon, G., & Backhouse, J. (2001). Current directions in IS security research: towards socio-organisational perspectives. *Information Systems Journal, 11,* pp. 127-153.

Dinnie, G. (1999). The Second annual Global Information Security Survey. *Information Management and Computer Security,* 7(3), pp. 112-120.

Drucker, P.F. (1988). The coming of the new organization. Harvard Business Review, Jan-Feb.

D.T.I. (2002), *Information Security Breaches Survey 2002*, Technical Report, April Department of Trade and Industry, London.

D.T.I. (2000), *Information Security Breaches Survey 2000*, Technical Report, April Department of Trade and Industry, London.

Ernst and Young, (2001), *Information security survey,* London: Ernst & Young.

Framel, J.E. Information value management. *Journal of Systems Management, 44*(12), 16.

Fulford, H., & Doherty, N.F. (2003). The application of information security policies in large UK-based organisations: An exploratory analysis. *Information Management and Computer Security, 11*(3), 106-114.

Gaston, S.J. (1996). *Information security: Strategies for successful management,* Toronto: CICA.

Gerber, M., von Solms, R., & Overbeek, P. (2001). Formalizing information security requirements. *Information Management and Computer Security, 9*(1), 32-37.

Glazer, R. (1993). Measuring the value of information: The information intensive organization. *IBM Systems Journal, 32*(1), 99-110.

Hair, J., Anderson, R., Tathem, R. & Black, W. (1998). *Multivariate Data Analysis*. Upper Saddle River, NJ: Simon & Schuster.

Haugen, S, & Selin, J.R. (1999). Identifying and controlling computer crime and employee fraud. *Industrial Management & Data Systems, 99*(8), 340-344.

Higgins, H.N. (1999). Corporate system security: Towards an integrated management approach. *Information Management and Computer Security, 7*(5), 217-222.

Hinde, S. (2002). Security surveys spring crop. *Computers and Security*, *21*(4), 310-321.

Hinde, S. (2003). Cyber-terrorism in context. *Computers and Security*, *22*(3), 188-192.

Hone, K., & Eloff, J.H.P. (2002). Information security policy- what do international security standards say? *Computers & Security*, *21*(5), 402-409.

Lindup, K.R. (1995). A new model for information security policies. *Computers & Security*, *14,* 691-695.

Loch, K.D., Carr, H.H., & Warkentin, M.E. (1992). Threats to information systems – Today's reality, yesterday's understanding. *MIS Quarterly*, *16*(2), 173-186.

McPherson, P. K. (1996). The inclusive value of information. *International Federation for Information and Documentation – 48th Congress*, Graz: pp. 41-60.

Menzies, R. (1993). Information systems security. In J. Peppard (Ed.), *IT Strategy for Business*, London: Pitman Publishing.

Mitchell, R.C., Marcella, R., & Baxter, G. (1999). Corporate information security. *New Library World*, *100*(1150), 213-277.

Moule, B., & Giavara, L. (1995). Policies, procedures and standards: An approach for implementation. *Information Management and Computer Security*, *3*(3),7-16.

Orna, E. (1999). '*Practical Information Policies*', Aldershot, Gower.

Pfleeger, C.P. (1997). *Security in computing*. Englewood Cliffs, NJ: Prentice Hall International.

Porter, M.E., & Millar, V.E. (1985). How information gives you competitive advantage. *Harvard Business Review*, Jul-Aug: pp149-160.

Post, G., & Kagan, A. (2000). Management trade-offs in anti-virus strategies. Information & Management, *37*(1), 13-24.

Premkumar, G., & King, W.R. (1992). An empirical assessment of information systems planning and the role of information systems in organisations. *Journal of Management Information Systems*, *19*(2), 99-125.

Romney, M. (1996). Reducing fraud losses. *New York State Society of Certified Public Accountants*, pp. 2-7.

Rostick, P. (1994). An Information Manifesto. *CIO*, Sept: pp. 24-26.

Schultz, E.E. (2002). A framework for understanding and predicting insider attacks. *Computers and Security, 21*(6), 526-531.

Siponen, M. (2000a). Policies for construction of information systems' security guidelines. *Proceedings of 15th International Information Security Conference (IFIP TC11/SEC2000),* Beijing, China, August 21-25, pp. 111-120.

Siponen, M.T. (2000b). A conceptual foundation for organizational information security awareness. *Information Management and Computer Security, 8*(1), 31-41.

Sterne, D.F. (1991). On the buzzword *'security policy'. Proceedings of the IEEE Symposium on Research in Security and Privacy, 219-230.*

Strassmann, P.A. (1985). Information payoff. New York, NY: The Free Press

Strong, D. M., Lee, Y. & von Solms, R. (1998). Information security management (1): Why information security is so important. *Information Management and Computer Security, 6*(5), 224-225.

von Solmes, R. (1998). Information security management (1): Why information security is so important. *Information Management & Computer Security, 6*(5), 224-225.

Wood, C.C. (1995). A policy for sending secret information over communications networks. *Information Management & Computer Security, 4*(3), 18-19.

Wood, C.C. (1996). Writing InfoSec Policies. *Computers & Security, 14*(8), 667-674.

# Chapter XIII

# Metrics Based Security Assessment

James E. Goldman
Purdue University, USA

Vaughn R. Christie
Purdue University, USA

## Abstract

*This chapter introduces the Metrics Based Security Assessment (MBSA) as a means of measuring an organization's information security maturity. It argues that the historical (i.e., first through third generations) approaches used to assess/ensure system security are not effective and thereby combines the strengths of two industry proven information security models, the ISO 17799 Standard and the Systems Security Engineering Capability Maturity Model (SSE-CMM), to overcome their inherent weaknesses. Furthermore, the authors trust that the use of information security metrics will enable information security practitioners to measure their information security efforts in a more consistent, reliable, and timely manner. Such a solution will allow a more reliable qualitative measurement of the return achieved through given information security investments. Ultimately, the MBSA will allow professionals an additional, more robust self-assessment tool in answering management questions similar to: "How secure are we?"*

# Introduction

Information security incidents are on the rise, with new attacks reported daily (for the latest statistics on system related incidents and security breaches refer to http://www.cert.org/stats/cert_stats.html). How have system administrators and security professionals reacted to these new threats? Historically, system owners have rushed to "acquire the latest cure" (Nielsen, 2000). They have tried in earnest to procure today's fix with little thought to the benefit truly gained from such utilities and or techniques. This approach to system security is changing; that is, the paradigm is shifting toward a model of increased accountability. Security managers are increasingly being held responsible for demonstrating the effectiveness of their security initiatives, for showing that their investments have provided not only value but also greater security to their respective organizations. In short, they are being asked, "How secure are we?" (Payne, 2001).

Answers to this and similar questions are not easy to derive (Payne, 2001). Dating back to the late 1970s and early 1980s when the annual loss expectancy (ALE) calculation was being developed by the Federal Information Processing Standard (FIPS), security professionals have attempted to define security by a single distinct value: ALE (Fletcher, 1995). Since that time, additional information security management tools and documents have been developed. In recent years, a number of guidance documents have been published to assist organizations, both in the public and private sectors, in establishing and maintaining their information technology security programs (Dr. Fletcher has described these guideline documents as third-generation information security tools). Examples of these documents include the NIST Handbook, the CSE Guide, BSI 7799, ISO 17799, and ISO/IEC 13335 (Hopkins, 1999). Unfortunately, problems reside in these guidance tools; specifically, their holistic nature makes it difficult to measure specific information security parameters easily, effectively or efficiently (Payne, 2001).

This chapter proposes a metric-based information security maturity framework constructed from the combination of the ISO 17799 Standard and the Systems Security Engineering Capability Maturity Model (SSE-CMM). While many believe the SSE-CMM to be simply another in the myriad of recently published best practices and general security guidelines (i.e., a supplement to the ISO 17799 Standard as opposed to its complement), their assessment is inaccurate (Hopkins, 1999). The MBSA will illustrate how the SSE-CMM can be used to measure the maturity of the information security practices implemented via the ISO 17799 Standard. The end result will be a self-facilitated metrics-based assessment model that enables organizations, from both the public and private sectors, to accurately assess the maturity of their information security processes. By using the SSE-CMM to measure the maturity of an organization's information security program (specifically the ISO 17799 Standard), the proposed solution will enable professionals to measure in a more consistent, reliable, and timely manner areas for improvement and effectiveness. Furthermore, it will allow a more reliable qualitative measurement of the return achieved through given information security investments. Ultimately, the solution offered in this chapter will allow professionals an additional, more robust self-assessment tool in answering: "How secure are we?"

# Statement of the Problem

Encouraged by a combination of the terrorist actions of September 2001, the mounting complexity of malicious online attacks, and the increasing realization that unbroken network surveillance, immediate intrusion detection and real-time response strategies are boardroom responsibilities, information security has come to the forefront of corporate and government agendas (Dargan, 2002). However, even with growing media attention and information technology (IT) spending predictions, most of the data seen by Ultima Business Solutions' security consultants suggest that more and more information technology teams are continuously failing in their responsibility to protect their organizations from attack (Dargan, 2002).

Emphasizing the importance of information security, the United States federal government is expecting to increase information security spending by a "compounded annual growth rate of 25%" between 2001 and 2006 (Mark, 2000); the private sector is expected to follow suit, where spending on e-security services will nearly triple from $4.3 B in 2000 to $11.9 B in 2005 (CIO.com, 2002). Even more impressive is the anticipated 28% increase in total security spending between 2000 and 2005 (from $8.7B to $30.3B) (CIO.com, 2001). However, with greater funding comes increased scrutiny, and rightly so. In a February 2002 article by Hurley, it was stated that even though security spending may be on the rise, companies still might not be more secure. Therefore, it is with good reason that information security managers are increasingly being asked to demonstrate a return on the investment(s) being made; for example, when they are asked (Payne, 2001):

- "Are we more secure today than we were before?"
- "How do we compare to our competition?"
- "Do these new security efforts result in more secure systems?"
- "If so, how do we know?"
- "What makes an effective security program?"
- "How are security programs measured?"
- "How secure are we?"

Executive management is truly attempting to uncover the value-added of their security spending. How will these questions be answered? Several approaches, both qualitative and quantitative, have been attempted. In recent years, guidance documents including the ISO 17799 Standard, the NIST Handbook, the CSE Guide, BSI 7799, ISO/IEC 13335, and so forth have evolved. These documents have attempted to qualitatively guide corporations in addressing these questions (Hopkins, 1999). While each framework differs from its peers (in terms of structure, culture and organization), each seeks to offer suggestions concerning applicable approaches to enterprise-wide security (Hopkins, 1999). That is, collectively, the various frameworks seek the common goals of explicitly documenting, in a single framework, the various facets of the system, such as the system's behavior, structure, and history (Craft, 1998). According to Dr. Fletcher (1995), these guidance documents seek to identify the system's purpose and behavior, structure, and relationship to its environment all in one common document (Craft, 1998). Unfortunately, with the exception of the ISO 17799 Standard, the structural, cultural and organizational differences make it difficult to transpose the various models across industries and or jurisdictions (i.e., private vs. public sector) (Hopkins, 1999). Furthermore, the information security industry has cited the following broad-level weaknesses associated with such frameworks.

- Their independence from actual risks may lead to:
    - either over-securing or under-securing information assets (or both);

- the lack of information collection on new threats and vulnerabilities; and

- difficulties in measuring the efficiency of security procedures (Chuvakin, 2002).

- They do little to measure the effectiveness of security investments (Payne, 2001); and

- They do not adequately answer those "difficult questions" being asked by executive management (Payne, 2001).

Carroll (2000) has proposed a potential solution to these known weaknesses: metrics. With metrics, a number of advantages are realized. These include:

- Processes become repeatable, more manageable; and therefore may be carried out more frequently on specific system segments;

- System security risk assessments can be performed immediately via functional templates that can potentially represent a number of similar functional systems;

- System targeting can be performed more frequently than current frameworks allow/recommend;

- Risk assessment processes and results between service providers may become more standardized;

- Threat, Vulnerability, Impact (T/V/I) baselines for similar functional systems may be created; and

- Planning, Programming & Budgeting System (PPBS) inputs for acquisition and development may realize improvements (Carroll, 2000).

Clearly, there are many benefits that may result from the definition of specific metrics in the field of information security (Payne, 2001). However, defining specific, timely metrics is not easy; in fact, as they pertain to the information security field, metrics are in their infancy (Nielsen, 2000). By finding a middle ground between the highly quantitative measures of the first generation and the qualitative measures and frameworks currently being used, we (as information security practitioners) will take a needed step toward the fourth generation of information security paradigms.

# Significance of the Problem

Though the lack of specific, timely measures with which to measure information security is significant, the overall requirement for information technology security is not questioned by organizations. Many businesses realize the need for secure information technology infrastructures, environments and systems, and furthermore understand that information assurance must be pervasive (Applied Computer Security Associates, 2001, p. 1). The public sector is in agreement. A review of United States federal government policies, guidelines and standards has revealed more than 15 distinct information security related documents since the mid 1980s. However, the need to measure and evaluate the effectiveness of the tools and techniques used to secure today's highly connected, always-on businesses continues to grow (Applied Computer Security Associates, 2001, p. 1); as documented by Neilson, this need has only recently been identified (p. 3). The desire to arrive at a single discrete value by which to buy or rate new technologies and/or to commit organizational resources to information security initiatives has largely been inadequate; in fact, as stated by the ACSA (2001), the techniques currently used are neither "generally accepted nor reliable measures for rating information technology security or requisite security assurance" (p. 1). Moreover, the risk assessment process, a quantitative or qualitative exercise, usually results in either a recommendation or a non-recommendation for a given solution, but fails to provide the needed Specific, Measurable, Attainable, Repeatable and Timely (SMART) metrics required to assess information security, let alone improve information system security (Carroll, 2000).

The underlying goal of all risk assessments is the identification of threats, vulnerabilities, and impacts (Carroll, 2000). In the ever-changing, fast-paced world of information technology, technology and operational requirements drive numerous hardware and software changes (Carroll, 2000). The result, due to the resource constraints of staff, money, and/or time, is that information technology systems commonly undergo changes not documented or assessed between the three year risk assessment cycles (Carroll, 2000). Without accurate information security data, how can information security staff reliably answer those questions being asked by management? How are they able to determine whether their efforts have/have not been successful?

As argued by the National Institute of Standards and Technology (NIST), information security metrics are needed to understand the current state of system security, to improve that state, and to procure/obtain the necessary resources to implement improvements (Nielsen, 2000). Further supporting the cause for metric program development is Carroll (2000), who has stated: "[d]ue to the complexity of variables in the risk assessment process, a structured approach with metrics is critical to a methodical assessment of goals, processes, procedures, and continuous improvements to the risk assessment process" (Carroll, 2000). The underlying premise of Hurley's February 2002 article was that of late, there has been no causal relationships between the amount of funding spent and the how secure an organization is. Effectively, there are no measures, no standard way of scoring security implementations (Hurley, 2002). Unfortunately, the practice of developing information security metrics is an undeveloped science; the MBSA is one of the first works in this field (Nielsen, 2000).

In short, metrics are useful to information/asset owners, information security risk assessment personnel, and information security engineers in evaluating design tradeoffs and security measures (Carroll, 2000). Metrics will assist information security engineers and staff in improving their processes, procedures, and risk assessment activities. They will provide the necessary information to acquire required resources, and give measures of performance (MOP) and measures of effectiveness (MOE) to determine information security maturity. Finally, metrics will make it possible to reliably answer those difficult questions being asked by executive management (Carroll, 2000; Payne, 2001).

As was earlier documented, recently published best practices and guiding frameworks have sought to provide a single source of information regarding the various facets of the system including its behavior, structure, and history; however, the structural, organizational and cultural differences (of all but the ISO 17799 Standard) make it difficult to transpose a given framework across industry borders (Hopkins, 1999). Furthermore, the primary weaknesses of these documents, including their inability to adequately measure the effectiveness of security investments (Payne, 2001), imply that it is exceedingly difficult to create SMART or effective metrics. Given that the underlying goal of the SSE-CMM is to measure the effectiveness of an organization's "Systems Security Engineering" processes, it (acting as a complement to these frameworks) can help alleviate the problem (Hopkins, 1999).

Specifically, using the SSE-CMM as the guiding framework and complementing that with the ISO 17799 Standard, a self-facilitated information security metrics model can be developed that will offer the security community a tool that may be transposed across cultural, organizational and structural jurisdictions. Furthermore, such a model will offer security professionals a flexible tool that can be adapted to their specific needs or easily used as a starting point in designing their own information security metrics. As a final contribution, a model of this nature may be used to better assess the maturity of an organization's information security practices, and provide a clearly defined path toward improvement.

# Purpose of the MBSA

With the problem firmly stated and the impact it represents to organizations known, the purpose then is to provide the information security profession a metrics-based security assessment (MBSA) that can measure the effectiveness and maturity of an organization's information security. The results of such an assessment can be used to justify current and future information security investments as well as answer, in a specific and timely manner, questions surrounding the measurement of organizational information security (i.e., "How secure are we?" and the like).

In developing the metrics-based security assessment (MBSA), we will first discuss the strengths and weaknesses associated with historical and traditional information security models, assessment techniques and paradigms. From this discussion, two proven and accepted models will be combined; specifically, the models are the ISO 17799 Standard and the Systems Security Engineering – Capability Maturity Model (SSE-CMM). Combining these models is necessary, as each have weaknesses that may yield negative impacts to organizations if not mitigated. Furthermore, the complementary nature of the models will build from each other's strengths and result in an information security assessment model that is applicable to a wide variety of audiences, including businesses that are:

- Public and/or private sector (regardless of industry and size)
- Profit and/or non-profit
- International as well as United States based organizations

Through the development of the MBSA, information security professionals will be better prepared to answer those questions asked by executive management; they will be better able to prove or disprove that the investments made in information security have resulted in a an environment comprised of greater data/information security. Furthermore, by gaining the ability to prove or disprove the impact(s) their investments have made on information security, information security professionals will be better prepared to justify current and future information security related expenses.

# Background

Of note are both the quantity and quality of reoccurring security related incidents in recent history. Specifically, many of the security incidents have resulted in one or more of the following: information system viral infections, Web page defacements, information theft, or denial of information technology service(s). And while these incidents, in and of themselves, are worthwhile avenues of study, they do little to represent the disturbing underlying situation: information security related incidents are increasing. Many times these incidents (Figure 1) affect tens of thousands of systems and cost real money (Baseline Magazine, 2002). Further illustrating this point is Streckler, a vice president at Symantec who works with government officials who, in April of 2002, made the comment, "we get about 15 new virus submissions every day" (Baseline Magazine, 2002).

*Figure 1. Number of incidents reported by year (CERT, 2003)*

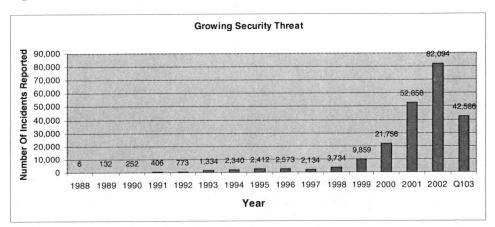

While the sheer increase in total reported incidents may be alarming, so too is the cost. For example, the Code Red virus infected 250,000 systems in under nine hours - the cost: an estimated $2.6 B (Baseline Magazine, 2002). Further research, stemming from 2001 data collected by Computer Economics, estimates United States' companies lost approximately $15 B due to virus infections alone in 2002 (Baseline Magazine, 2002). And while the impact of these examples has been felt hard, so to has the resulting downtime, costing United States' businesses (in the 12 months prior to September 11, 2001) nearly $273 M, and $1.39 T worldwide (Bruck, 2002).

Certainly, the inability to measure the level of information security has been costly; so much so that a significant effort within the information security profession (since the early 1970s) has been focused on addressing it. The following sections illustrate the strengths and weaknesses of the historical and traditional methodologies.

# Methodology

With a high-level history of information security related incidents, the question becomes, "How can we hope to overcome and/or withstand the security related assault(s) being taken on our systems?" One method, seemingly popular by both private and public sectors, is an increase in security spending. In a 2001 study conducted by Gartner Group, security spending was predicted to increase tenfold by the year 2011 (Lemos, 2001). This trend is certainly welcome by security professionals, as it gives reason to believe that greater emphasis will be placed on overcoming information security problems. However, a shift in focus lies in its wake (Payne, 2001). As stated by Payne (2001) of the SANS Institute, under these circumstances, senior management will adjust its current viewpoint of information security as a necessary evil (Girt, 2002), to an investment like any other, accountable for demonstrating a timely return (Payne, 2001).

Since the 1970s, three generations of information security management/ measurement have been used to demonstrate return on security investment (ROSI) (Carroll, 2000). The first two generations of information security measurement were largely built off of an underlying Annual Loss Expectancy (ALE) algorithm. The basic premise of the ALE calculation is that information security professionals can calculate a single discrete value of information security. Though the discrete value derived through combining risk, frequency, and impact led to ALE's widespread acceptance, it also is regarded as a principal cause of problems (Gilbert, 1989). Specifically, the lack of empirical data regarding frequency of specific risks and the impact those risks cause has left security officials at a loss for information (Gilbert, 1989).

More recently, a number of frameworks aimed at guiding information security implementation and maintenance programs have been developed (Hopkins, 1999). Examples include the International Organization for Standardization (ISO) 17799 Standard, the National Institute of Standards and Technology (NIST) Handbook, the Communications Security Establishment (CSE) Guide, British Standards Institute (BSI) 7799, and the ISO/IEC 13335 (Hopkins, 1999). While each document uses various structures and organization, they each attempt to provide assistance related to critical areas of concern that should be addressed by an information security program (Hopkins, 1999). As can be expected, the structure and organization of each framework is that which is meaningful for the community for whom and by whom they were developed (Hopkins, 1999). Furthermore, each framework offers suggestions concerning techniques and approaches for framework compliance, but they also under-stand the need for customized guidance — each organization is expected to have differing needs, goals, values, and so forth, and will want to customize the frameworks to meet their environment, goals, vision, and/or culture (Hopkins, 1999). Though the current state of information security measurement and management lies in these third-generation models, problems must still be overcome. Specifically, the security frameworks put in place do little to measure the effectiveness of security investments; they do not adequately answer those difficult questions being asked by executive management (Payne, 2001).

Serving to identify their strengths, weaknesses and complementary nature, the coming sections introduce both the ISO 17799 and SSE-CMM methodologies.

## The ISO 17799 Standard

ISO 17799 is an internationally recognized Information Security Management Standard, first published by the International Organization for Standardization, or ISO (http://www.iso.ch), in December 2000. The standard itself meets the needs of the MBSA in the following ways: it (the ISO 17799) is high level, broad in scope, and conceptual in nature (Carlson, 2001). This approach allows it to be applied across a variety of enterprises and applications; however, this generalist approach has also made the standard controversial among those who believe standards should be more precise (Carlson, 2001). In spite of this controversy, ISO 17799 is the only "standard" devoted to Information Security Management in a field generally governed by "guidelines" and "best practices" (Carlson, 2001).

ISO 17799 defines information as an asset that may exist in various forms and has value to an organization (Carlson, 2001). The goal of information security is to suitably protect this asset in order to ensure business continuity, minimize business damage, and maximize return on investments. As defined by ISO 17799, information security is characterized as the preservation of confidentiality, integrity and availability (Carlson, 2001). With these definitions in mind, it is important to gain an understanding of what ISO 17799 is not; in short, the standard is not (Carlson, 2001):

- Technical in nature;
- Product or technology driven;
- An evaluation methodology such as the Common Criteria/ISO 15408 (http://www.commoncriteria.org), which concerns itself with functional and assurance requirements of specific equipment;
- Associated with the "Generally Accepted System Security Principles" (GASSP) (http://web.mit.edu/security/www/gassp1.html), which is a collection of information security best practices; nor is it
- Correlated to the five-part "Guidelines for the Management of IT Security" (GMITS/ISO 13335), which provides a logical framework for managing information technology security.

The ISO 17799 is in fact a descendant of the British Standard Institute (BSI) Information Security Management standard BS 7799 (http://www.bsi-global.com), which has long been proactive in the evolving arena of information security (Carlson, 2001). In 1993, a working group devoted to information security was developed to effectively create/construct a "Code of Practice for Information Security Management" (Carlson, 2001). The completed work then evolved into the first version of the BS 7799 standard released in 1995 (Carlson, 2001). In 1998 and again in 1999, in response to industry demands, the BSI created, as a component to the BS 7799, an accreditation program for auditing firms known as "Certification Bodies," which is known today as c:cure (http://www.c-cure.org) (Carlson, 2001). Today, the BS 7799 standard consists of two parts, Part 1: Code of Practice, and Part 2: Specification of Information Security Management Systems (Carlson, 2001).

While some organizations have utilized the BS 7799 standard, demand for an internationally recognized information security standard, under the guidance of an internationally recognized body such as the ISO, has matured (Carlson, 2001). This pressure led to the "fast tracking" of BS 7799 Part 1 (by BSI) and resulted in the ISO's first release of the standard ISO/IEC 17799:2000, in December 2000. As of September 2001, only BS 7799 Part 1 has been accepted for ISO standardization, because it is applicable internationally and across all types of organizations (Carlson, 2001). To date, efforts to submit BS 7799 Part 2 for ISO standardization have been halted (Carlson, 2001).

With a background of ISO 17799 in hand, a more in-depth overview of the standards contents is in order. As a precursor to the actual elements within the standard, it is important to recognize that the controls developed and documented within the ISO 17799 standard may not be relevant to every situation (BS ISO/IEC 17799:2000, 2001). That is, the standard cannot take into account "local system, environmental, or technological constraints" and may not be in a form that suits every possible user or organization (BS ISO/IEC 17799:2000, 2001). With this in mind, the ISO 17799 standard and the resulting MBSA may need to be supplemented by further guidance (e.g., HIPPA, SANS, CERT, CERIAS, etc., documentation) (BS ISO/IEC 17799:2000, 2001).

Once the organization has determined the applicability of the ISO 17799 standard to its specific business requirements, appropriate controls can then be put in place to help mitigate or offset the risks identified, effectively reducing the impact felt in those environments (Carlson, 2001). ISO 17799 consists of ten security controls, which are used as the basis for the security risk assessment (Carlson, 2001, p. 6). Table 1 details the ISO 17799 Security Controls as defined by Carlson (2001).

Though the ISO 17799 is indeed a standard, it has come under pressure. Critics have described the standard as being too flexible, too open, and too loosely structured to provide any significant value (Walsh, 2002). They have used phrases such as a "mile wide and an inch deep" to refer to its cross-industry applicability and its "best-practice" nature (Walsh, 2002). In fact, some go so

*Table 1. ISO 17799 security control definitions*

| Security Control | Definition |
|---|---|
| Security Policy | Security Policy control addresses management support, commitment, and direction in accomplishing information security goals; it is composed of Information Security Policy, and Ownership and Review. |
| Organizational Security | Organizational Security control addresses the need for a management framework that creates, sustains, and manages the security infrastructure; it is composed of Management Information Security Forum, Information System Security Officer, Information Security Responsibilities, Authorization Processes, Specialist Information, Organizational Cooperation, Independent Review, Third-Party Access, and Outsourcing. |
| Asset Classification and Control | Asset Classification and Control addresses the ability of the security infrastructure to protect organizational assets; it is composed of Accountability and Inventory, Classification, Labeling, and Handling. |
| Personnel Security | Personnel Security control addresses an organization's ability to mitigate risk inherent in human interactions; it is composed of Personnel Screening, Security Responsibilities, Terms and Conditions of Employment, Training, and Recourse. |
| Physical and Environmental Security | Physical and Environmental Security control addresses risk inherent to organizational premises; it is composed of Location, Physical Security Perimeter, Access Control, Equipment, Asset Transfer, and General. |
| Communications and Operations Management | Communication and Operations Management control addresses an organization's ability to ensure correct and secure operation of its assets; it is composed of Operational Procedures, Change Control, Incident Management, Segregation of Duties, Capacity Planning, System Acceptance, Malicious Code, Housekeeping, Network Management, Media Handling, and Information Exchange. |
| Access Control | Access Control addresses an organization's ability to control access to assets based on business and security requirements; it is composed of Business Requirements, User Management, User Responsibilities, Network Access Control, Host Access Control, Application Access Control, Access Monitoring, and Mobile Computing. |
| System Development and Maintenance | System Development and Maintenance control addresses an organization's ability to ensure that appropriate information system security controls are both incorporated and maintained; it is composed of System Security Requirements, Application Security Requirements, Cryptography, System Integrity, and Development Security. |
| Business Continuity Management | Business Continuity Management control addresses an organization's ability to counteract interruptions to normal operations; it is composed of Business Continuity Planning, Business Continuity Testing, and Business Continuity Maintenance. |
| Compliance | Compliance control addresses an organization's ability to remain in compliance with regulatory, statutory, contractual, and security requirements; it is composed of Legal Requirements, Technical Requirements, and System Audits. |

far as to say that implementing the standard may give the organization a false sense of security (Walsh, 2002). Walsh (2002) has described the ISO 17799's approach to information security in the following manner:

*ISO 17799's broad-brush approach may make it universally adaptable, but the standard can hardly stand by itself. Even as ISO works to amend the document, the standard will continue to rely on more specific security standards to buttresses its framework.* (Walsh, 2002)

With this said, the SSE-CMM may very well prove to be that "more specific" documentation to support the ISO 17799. In terms of measurement and the ability to qualitatively assess the information security investments at a given organization, the ISO 17799 is also lacking. As Walsh (2002) has stated, the ISO 17799 seldom attempts to provide guidance in assessing or understanding information security measures; the standard is short on mechanics for measuring the standard's effectiveness when put into practice (Walsh, 2002). By leveraging the SSE-CMM, these weaknesses can be overcome. The following section describes the SSE-CMM and shows how it may be used to further shore up the ISO 17799's shortcomings.

## Systems Security Engineering Capability Maturity Model

Building off the reputation of the SW-CMM (Capability Maturity Model for Software), the SEI (Software Engineering Institute) created the Systems Security Engineering Capability Maturity Model (SSE-CMM) with the goal of providing a model useful in assessing the level of security maturity surrounding an organization's systems (Hopkins, 1999). The underlying principle of the SSE-CMM is that, with all things being equal, the more mature the process, the better, more consistent the outcomes of that process will be. This is independent of the specifics, approach or methodology used within that process (Hopkins, 1999). That is to say, the SSE-CMM was specifically engineered to be approach and methodology neutral, implying that it could easily complement any of the methodologies previously evaluated. While the SSE-CMM can certainly be used to complement a given framework, there would need to be sound reasoning behind such an action. As stated by the SSE-CMM organization, the implementation of this maturity model "will maximize synergy between life cycle phases, minimize effort within each phase, eliminate duplication of effort, and result in more secure solutions, with greater assurance and at much lower cost" (SSE-CMM, 2003).

The foundation from which the SSE-CMM was constructed is a register of defined Process Areas (PAs), which are further broken down into more easily measured Best Practices (BPs) that are effectively used to determine the level of maturity assessed. One of the unique facets of the SSE-CMM is the inherent customization found within the model:

> [i]n the development of the SSE-CMM Model the problem was recognized that different organizations were likely to be using different combinations of Base Practices and Process Areas to achieve their objectives. Thus the Model was designed such that the organization could restructure the SSE-CMM Process Areas to fit their particular situation. (Hopkins, 1999)

That is to say, the "how of the BP" is not as critical as the process itself. There are a variety of methods that may be employed to achieve a particular objective/ activity; rather, what is important are the objectives of the activity, the processes and their associated outcomes (Hopkins, 1999). The PAs are the focus of the SSE-CMM, and as such are significant to the MBSA. The following paragraphs will provide a high level overview of the Security Engineering PAs found within the Security Base Practices chapter of the SSE-CMM (1999). It should be noted that the SSE-CMM is broken into two core components: Security Engineering PAs and Project PAs and Organizational PAs. Note only the Security Engineering PAs (Table 2) are leveraged in the MBSA. The goals identified within Table 2 have been provided by Hopkins (1999).

Similar to the SW-CMM, the SSE-CMM defines six levels of system security maturity, each of which is defined in Figure 2, where each progressive level represents a greater level of system security maturity (i.e., Level 5 represents a higher maturity than Level 3).

With an understanding of both the SSE-CMM and the ISO 17799 standard, it is important to reiterate that the SSE-CMM can easily complement an underlying information security model (such as the ISO 17799 standard) (Hopkins, 1999). In fact, some have criticized the SSE-CMM, stating that to truly assess the security of an organization's systems, the SSE-CMM should not stand alone; it needs an underlying methodology (Hopkins, 1999). We have already identified areas of weakness in the ISO 17799 standard, and now we see how the SSE-CMM may be used to help overcome those limitations; effectively, allowing security professionals an opportunity to measure the

## Table 2. SSE-CMM security engineering process areas

| Process Area | Description |
|---|---|
| PA 01 | **Administrator Security Controls** <br> Summary: The Administrator Security Controls PA is concerned with establishing security responsibilities, managing security configurations, awareness, training, and education programs, as well as security services and control mechanisms (SSE-CMM, 1997). <br> Goals: Security controls are properly configured and used. |
| PA 02 | **Assess Operational Security Risk** <br> Summary: The Assess Operational Security Risk PA is concerned with selecting a risk analysis method, prioritizing operational capabilities and assets, identifying threats, and assessing operational impacts (SSE-CMM, 1997). <br> Goals: The security impacts of risks to the system are identified and characterized. |
| PA 03 | **Attack Security** <br> Summary: The Attack Security PA is concerned with scope attack, developing attack scenarios, performing attacks, and processing attack results (SSE-CMM, 1997, p. 2-40) <br> Goals: An understanding of the security risk associated with operating the system within a defined environment is achieved. Risks are prioritized according to a defined methodology. |
| PA 04 | **Build Assurance Argument** <br> Summary: The Build Assurance Argument PA is concerned with identifying assurance objectives, defining an assurance strategy, controlling assurance evidence, analyzing assurance evidence, and providing assurance arguments (SSE-CMM, 1997). <br> Goals: The work products and processes clearly provide the evidence that the customer's security needs have been met. |
| PA 05 | **Coordinate Security** <br> Summary: The Coordinate Security PA is concerned with defining security objectives, identifying coordination mechanisms, facilitating coordination, and coordination of security decisions/recommendations (SSE-CMM, 1997). <br> Goals: All members of the project team are aware of and involved with security engineering activities to the extent necessary to perform their functions. Decisions and recommendations related to security are communicated and coordinated. |
| PA 06 | **Determine Security Vulnerabilities** <br> Summary: The Determine Security Vulnerabilities PA is concerned with selecting vulnerability analysis methods, analyzing system assets, identifying threats and vulnerabilities, and processing system vulnerabilities (SSE-CMM, 1997). <br> Goals: An understanding of system security vulnerabilities within a defined environment is achieved. |
| PA 07 | **Monitor Security Posture** <br> Summary: The Monitor Security Posture PA is concerned with analyzing even records, monitoring changes, identifying security incidents, monitoring safeguards, reviewing security posture, managing security incident responses, and protecting security monitoring archives (SSE-CMM, 1997). <br> Goals: Both internal and external security related events are detected and tracked. Incidents are responded to in accordance with policy. Changes to the operational security posture are identified and handled in accordance with the security objectives. |
| PA 08 | **Provide Security Input** <br> Summary: The Provide Security Input PA is concerned with understanding security input requirements, determining limitations and constraints, identifying alternatives, analyzing security engineering alternatives, providing security engineering guidance, and offering operational security guidance (SSE-CMM, 1997). <br> Goals: All system issues are reviewed for security implications and are resolved in accordance with security goals. All members of the project team have an understanding of security so they can perform their functions. The solution reflects the security input provided. |
| PA 09 | **Specify Security Needs** <br> Summary: The Specify Security Needs PA is concerned with gaining an understanding of security needs, identifying applicable laws, policies, standards and limitations, recognizing the system security context, defining security related requirements, and obtaining agreement on security initiatives (SSE-CMM, 1997). <br> Goals: A common understanding of security needs is reached between all parties, including the customer. |
| PA 10 | **Verify and Validate Security** <br> Summary: The Verify and Validate Security PA is concerned with identifying verification and validation targets, defining verification and validation approach, performing validation and verification, and providing validation and verification results (SSE-CMM, 1997). <br> Goals: Solutions meet security requirements. Solutions meet the customer's operational security needs. |

*Figure 2. SSE-CMM levels of maturity (SSE-CMM, 1997)*

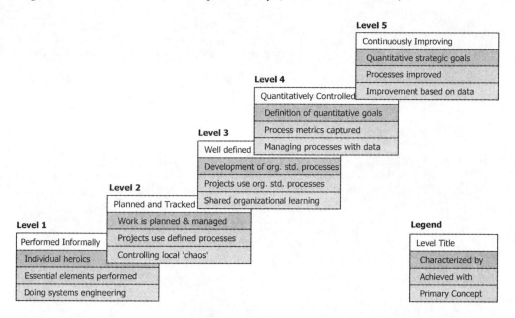

maturity (or effectiveness) of their security guideline (in this case, the ISO 17799). As a recommendation, future work should attempt to cost justify or measure the dollar value of this research in a similar manner to that used to cost justify the SW-CMM. This would be a worthwhile goal, as many cost justifications have increasingly been required for information security investments (Payne, 2001).

With an understanding of both the ISO standard and the SSE-CMM, it is now possible to begin work on actually merging the two documents and forming the metrics-based assessment (MBSA) model.

## MBSA Introduction

The MBSA allows the organization to rate itself on the set of baseline security controls that have been determined to coincide with both the ISO 17799 standard and the SSE-CMM. An initial pass through the questionnaire will result in the organization's baseline score from which additional assessments can/must be made. It is from these data that the organization will be able to determine the effectiveness of their information security initiatives and thereby finally answer the question: "How secure are we?" (Craft, 1998).

The following section offers a high-level review of the SSE-CMM levels of maturity. After completing the baseline assessment, organizations should identify appropriate goals regarding the maturity level to attain and proceed with information security projects/processes in such a way that those goals may be met in an appropriate (appropriate to the organization and its business drivers) time frame. Note that as a delimitation of this research, the author has not determined the necessary processes to achieve specific levels of information security maturity; rather, this task is left to the organization and may be facilitated by reviewing industry accepted best practices from SANS, CERT, CERIAS, Microsoft, RSA, and so forth. This is illustrated, at a high level in Figure 3 and at a more granular level in Figure 4.

At a more comprehensive level, Figure 4 illustrates a proposed model mapping the MBSA to the SSE-CMM process. Within the "Identify current and desired state" step, the organization would identify its "current state" information security maturity via the MBSA process. From there, a "desired state" of information security maturity can be determined (i.e., maturity level goals) via business drivers, benchmarking, and so forth.

*Figure 3. MBSA architecture*

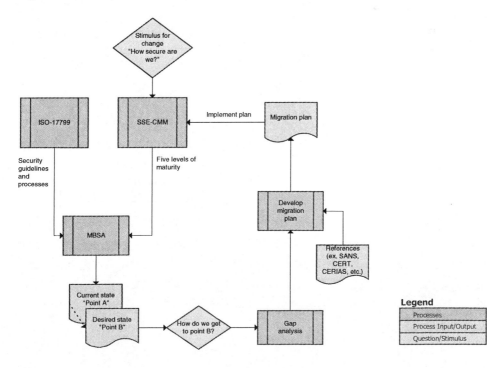

*Figure 4. MBSA to SSE-CMM process model mapping*

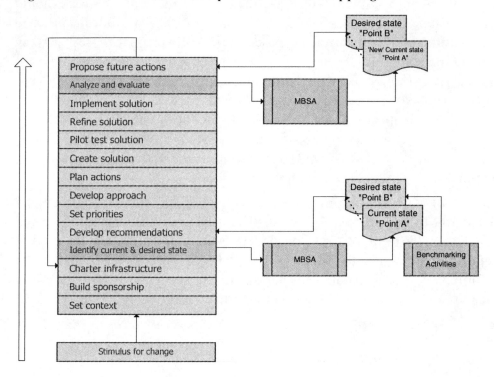

Progressing further upward through the SSE-CMM model, the organization would then reach the step of "Analysis and evaluation," where it should conduct the MBSA once again, assess the results/findings (i.e., the "New current state" after implementing the migration plan) against the "desired state" maturity level defined in the initial stages of the process model and identify potential future actions, resulting in an iterative approach to information security maturity. Note that conducting the full MBSA may not be required; depending on the business drivers at hand, the organization may choose only to assess the changes that were implemented and identify the level of maturity attained by such projects.

An underlying requirement of the MBSA is that of a self-assessment; that is to say that the group responsible for the audit must make a realistic assessment and rate themselves and their organization on the maturity of their security practices. The following section defines a response scale that could easily be leveraged to meet this requirement.

# Response Scale

A sample MBSA template is illustrated in Table 3; a resulting sample metric is illustrated in Table 5.

For each baseline control, a response (on a 0 to 5 scale) indicating the degree to which the control has been implemented should be recorded. The two extreme scores are 0 and 5:

- Score a baseline control as 0 if the baseline control is required but has not been implemented in the organizational entity for which responses are being sought and there has been no effort put forth that might ultimately lead to implementation. For example, if the baseline control is the development and promulgation of a security policy, score the baseline as 0 if there is currently no security policy, none is in development, and there are no plans to develop one.

- If the control is not required or the question is not applicable to the organization, score it as N/A.

- Score a baseline control as 5 if the baseline control has been fully implemented in the organizational entity for which responses are being sought and the assessor is satisfied with the quality and completeness of that implementation. For example, if the baseline control is the development and promulgation of a security policy, score the baseline control as a 5 if there is:

  - currently a fully developed and published security policy, and
  - the assessor is fully satisfied with the quality and completeness of the security policy and its implementation throughout the entity.

Generally speaking, values between 0 and 5 should reflect the extent of implementation. For example, if the security policy is 20% of the way towards level 5 maturity, score the control as a 1. If the security policy is 60% of the way towards level 5 maturity, score the control as a 3.

Scores can be influenced by varying degrees of implementation within the organizational entity. If one part of the entity has completely implemented a security policy, and another part of the entity has rejected that policy and has no plans to develop their own, the control should by scored as a 2 (rounding down to limit the effects of "false security") for the entire organization.

Assigning scores to controls is most straightforward if they are thought of in the following manner: the score of 0 as being 0% of the way towards full and complete attainment of maturity level goals and 5 as being 100%. Scores between 0 and 5 signify only partial implementation of the ideal maturity level (level 5). Table 3 represents the format used for each metric.

After assessing/scoring each metric, the assessor can calculate the average obtained. The average for each process area may be reported (to management)

*Table 3. Metric template*

| Category | Description |
|---|---|
| Metric | Identify the metric being assessed. |
| Maturity Level Goal | Identify the SSE-CMM level goal for this control (e.g., Level 2, etc.). |
| Scale/Rating | Identify the percentage of maturity level goal attainment via the previously documented definitions (see Response Scale, above). |
| Frequency | Identify:<br>　The date of the current assessment.<br>　An appropriate frequency of assessments (i.e., how often).<br>　Proposed date for the next assessment. |
| Implementation Evidence | Justify the assessor's rating of the control. |
| Data Source | Identify the sources/references/resources used to make the assessment. |

*Table 4. Legend for yearly maturity assessment*

| Legend |
|---|
| 2002 – <u>Underlined</u> |
| 2003 – Highlighted |
| 2004 – Box |
| 2005 – **Bold** |

*Table 5. Sample metric*

| Metric | Terms and conditions of employment; state employee responsibilities regarding information security (where appropriate) are continued for a period of time after the employment period. |
|---|---|
| Maturity Level Goal | Level 3 |
| Scale/Rating | 0—1—2—3—4—5   N/A   Unknown |
| Frequency | Current Date:<br>Frequency:<br>Next Assessment Date: |
| Implementation Evidence | Sample only. |
| Data Source | Sample only. |

and therefore readily identify the level of maturity for each process area within the SSE-CMM. Should the organization elect to assess each process area on an annual basis, the following legend may prove useful; it allows the organization to easily indicate up to four years of maturity within the assessment document.

For instance, in PA01, the assessor could quickly identify the levels of maturity found for 2002 through 2005. Similarly, the same legend can be used for each individual metric, allowing the assessor to clearly identify the maturity levels attained in previously conducted MBSAs. For example, Table 5 indicates that in 2002 and 2003, the organization or department was only at Level 1 for the defined sample metric. In 2004, the system progressed to Level 2; and in 2005, the goal of Level 3 maturity was attained.

# Conclusion

Historically, organizations have sought to measure information security using such methods as the annual loss expectancy (ALE) calculation or more recently developed third-generation information security management frameworks. However, problems lie in these measurement techniques. For instance, the ALE calculation assumes a fixed information security threat environment and known values for the probability of a given information security vulnerability being exploited and the known cost of such a breach in information security. This is not the case, as organizations have had difficulty valuing the intangible assets at risk (i.e., customer loyalty, company image, etc.). Furthermore, identifying an accurate probability/likelihood of a given threat exploiting a particular vulnerability is still largely a mystery.

In terms of more recent, third-generation information security models, the significant problems rest in their inability to allow organizations to measure and assess the levels of information security within the organization. Furthermore, these methodologies lack the ability to demonstrate a progression toward organizational information security goals. Though weaknesses exist in these models, they do offer the significant advantage of providing the necessary processes that should be measured. What is needed is a means in which to measure them.

In order to overcome the weaknesses associated with existing methods, this chapter proposes a solution that leverages an internationally accepted/standardized framework: the ISO 17799 Standard. In doing so, relevant measures of importance that have been proven significant to industry will be used to create the metrics-based security assessment proposed in this chapter. Furthermore, using a proven measurement model, the SSE-CMM, the solution identified in this chapter adds an industry accepted measurement techniques to the ISO 17799 Standard, thereby overcoming its primary weaknesses.

By building an MBSA aligned with these proven information security guidelines, a solution offering a broad range of applicability (i.e., the ability to generalize/apply), reaching across a considerable spectrum of industries, organizations and industrial sectors is achieved. Furthermore, specific information security metrics will not only make those difficult to answer questions be more easily managed, but they will enhance system security, allow a more efficient use of time and staff, and make providing technical support less cumbersome (i.e., once metrics are in place, requiring all systems comply to them is expected to help ease technical support issues) (Livingston, 2000).

Finally, information security metrics are expected to provide an easily understandable snapshot of the organization's "current state" (i.e., current level of information security maturity), and provide a roadmap by which to continuously improve. The metrics-based security assessment proposed in this chapter will assist information security professionals as they attempt to obtain the necessary resources to implement improvements, and to evaluate the trade-offs intrinsic in security mitigating measures.

# References

Alldas. (2003). Alldas.de – IT Security Information Network [Online]. Available: http://www.security.nl/misc/alldas.html

Applied Computer Security Associates. (2001). *Proceedings: Workshop on Information Security System Scoring and Ranking.* Applied Computer Security Associates and the MITRE Corporation.

Baseline Magazine. (2002, April). Disaster prevention strategies. *Ziff Davis Media, 005,* 3a.

Blair, R.C. (2001, April 12). *What is the cost? What is the return on investment?* Pittsburgh PMI Chapter.

Bruck, M. (2002, February). *The business benefits of adequate information security. Cisco World Magazine* [Online]. Available:http://www.ciscoworldmagazine.com/monthly/2002/02/securityroi.shtml

BS ISO/IEC 17799:2000. (2001, June). Information technology – Code of practice for information security management. *British Standard.*

Canada (2001, December 11). Canadian handbook on information technology security (MG-9). *Communications Security Establishment* [Online]. Available:http://www.cse.dnd.ca/en/knowledge_centre/publications/manuals/MG-9.html

Carlson, T. (2001, September 25). *Information security management: Understanding ISO 17799.* Lucent Technologies.

Carroll, J.M. (2000, July 6). *A metrics-based approach to certification and accreditation.* BTG Inc.

CERT. (2003, July 4). *CERT Coordination Center.* CERT/CC. Available: http://www.cert.org/stats/

Chuvakin, A. (2002, January 28). Approaches to choosing the strength of your security measures. *LinuxSecurity.com* [Online]. Available:http://www.linuxsecurity.com

CIO.com Metrics. (2001, January 25). Behind the numbers. *CIO.com* [Online]. Available: http://www2.cio.com/metrics/index.cfm?CATEGORY=29&Go=Go

CIO.com Metrics (2002). Behind the numbers. *CIO.com* [Online]. Available: http://www2.cio.com/metrics/index.cfm?CATEGORY=29&Go=Go

Craft, R. et al. (1998, August 6). *An open framework for risk management.* National Institute of Standards and Technology. Available: *http://csrc.nist.gov/nissc/1998/proceedings/paperE6.pdf*

Dargan, L. (2002, August 24). Smashing the milestone. *SC Info Security Magazine* [Online]. Available:http://www.scmagazine.com/scmagazine/sc-online/2002/article/32/article.html

Defacement Commentary Mail List Archive. (2003). [Online]. Available:http://www.attrition.org/security/commentary/

Fletcher, S., Jansma, R., Lim, J., Halbgewachs, R. Murphy, M., & Wyss, G. (1995). Software system risk management and assurance. *Proceedings of the 1995 New Security Paradigms Workshop,* August 22-25, San Diego, CA.

Gilbert, I. (1989, October). *SP174 guide for selecting automated risk analysis tools.* National Institute of Standards and Technology.

Girt, K. (2000, December 14). *How to hit the jackpot with a security investment.* Applied Computer Security Associates Presentation.

Goldman, J., & Christie, V. (2003). Measuring information security: Combining the SSE-CMM with the ISO 17799 Standard. *Proceedings of the 2003 Information Resource Management Association International Conference,* May 18-21, Philadelphia, PA.

Hopkins, J.P. (1999). *The relationship between the SSE-CMM and IT security guidance documentation,* 1-7. EWA-Canada Ltd.

Hurley, E. (2002, February). Security spending may be up, but does that mean more security? SearchSecurity.com [Online]. Available: *http://searchsecurity.techtarget.com/originalContent/0,289142,sid14_gci802865,00.html*

Initiative. (2001, December 20). *A comparative study of IT security criteria. Initiative D21.* Available: http://www.initiatived21.de/arbeitsgruppen/5sicherheit/leitfaden-e.pdf

ITSEC. (1997). *About the ITSEC scheme.* Available: http://www.itsec.gov.uk/info/about.htm#T1

Jazri, H. (2001, April 16). *The regulatory environment in information security.* SANS Institute. Available: http://rr.sans.org/legal/regulatory.php

Jenkins, J. (2001, August 21). *Organizational IT security theory and practice: And never the twain shall meet?* The SANS Institute. Available: http://www.sans.org/rr/papers/48/448.pdf

Lemos, R. (2001, June 11). *Security spending to jump tenfold by 2011* [Online]. Available: http://news.com.com/2100-1001-268209.html?legacy=cnet

Lemos, R. (2003, January 31). Counting the cost of slammer [Online]. Available: http://news.com.com/2100-1001-982955.html

Mark, R. (2000, March 13). U.S. to triple information security spending. *Security Planet* [Online]. Available: http://www.esecurityplanet.com/trends/article/0,,10751_991391,00.html

Micksch, A. (2000, September 13). *Information systems risk analysis, assessment and management.* The SANS Institute. Available: http://rr.sans.org/policy/risk.php

National Institute of Standards and Technology. (1981). *Proceedings of the Computer Security Risk Management Model Builders Workshop* (Washington, DC: National Institute of Standards and Technology, 1981).

National Institute of Standards and Technology. (1994). *FIPS 191. Guideline for the Analysis of Local Area Network Security*. http://www.itl.nist.gov/fipspubs/fip191.htm

National Institute of Standards and Technology. (1996). *Special Publication 800-12 An Introduction to Computer Security*: http://csrc.nist.gov/publications/nistpubs/800-12/

National Institute of Standards and Technology. (2000). *Federal Information Technology Security Assessment Framework*. NIST. November 28, 2000.

National Institute of Standards and Technology. (2003). *Computer Security Division – Mission* [Online]. Available: http://csrc.nist.gov/mission.html

Nielsen, F. (2000). *Approaches to Security Metrics*. http://csrc.nist.gov/csspab/june13-15/metrics_report.pdf

NISS. (2000, September). *National Information Systems Security (INFOSEC) Glossary*. NSTISSI. (May 7, 2002).

OWL. (2003). Evaluating Sources: Bibliographic Citations. Retrieved on March 15, 2002.

Paller (2003, January, 25). SQL Slammer Worm. http://www.sans.org/news-letters/newsbites/vol5_4.php

Payne, S. (2001, July 11). *A Guide to Security Metrics*. SANS Institute.: http://www.sans.org/rr/paper.php?id=55

PriceWaterhouseCoopers. (2000). *American Society for Industrial Security/PricewaterhouseCoopers Trends in Proprietary Information Loss Survey Report*. PriceWaterhouseCoopers. http://www.pwcglobal.com/extweb/ncsurvres.nsf/DocID/36951F0F6E3C1F9E852567FD006348C5

Robinson, S. (2003). Corporate espionage 101. http://www.sans.org/rr/paper.php?id=512

SANS. (2003). *What is the SANS Institute?* SANS Institute. http://www.sans.org/aboutsans.php

Schwalbe, K. (2002). *Information Technology Project Management (2nd ed)*. Canada: Thompson Learning.

SSE-CMM. (1997, June 16). *Systems Security Engineering Capability Maturity Model*. Carnegie Mellon.

SSE-CMM. (1999, April 1). *Systems Security Engineering Capability Maturity Model Version 2.0*. Carnegie Mellon.

SSE-CMM. (2002, July 9). *The Vision*. SSE-CMM. http://www.sse-cmm.org/vision.html

SW-CMM. (2003, July 10). *About the SEI-Welcome*. Carnegie Mellon Software Engineering Institute.: http://www.sei.cmu.edu/about/about.html

United States General Accounting Office. (1999). *Information Security Risk Assessment: Practices of Leading Organizations*. http://www.gao.gov/special.pubs/ai00033.pdf

United States General Accounting Office. (2000). *U.S. Govt. was disrupted by Computer Virus, GAO Says*. July 12, 2003, from: http://usinfo.state.gov/topical/global/ecom/00051902.htm

United States General Accounting Office. (2001). *Federal Information Security Controls Audit Manual*. http://www.gao.gov/special.pubs/ai12.19.6.pdf

Walsh, L. (2002, Marsh). Security Standards. http://www.infosecuritymag.com/2002/mar/iso17799.shtml

Chapter XIV

# The Critical Role of Digital Rights Management Processes in the Context of the Digital Media Management Value Chain

Margherita Pagani
Bocconi University, Italy

## Abstract

*This chapter sets out to analyze the impact generated by the adoption of Digital Rights Management (DRM) processes on the typical Digital Media Management Value Chain activities and tries to analyze the processes in the context of the business model. Given the early stage of the theory development in the field of DRM, the study follows the logic of grounded theory (Glaser & Strauss, 1967) by building the research on a multiple-case study methodology (Eisenhardt, 1989). The companies selected are successful players that have adopted DRM processes. These companies are Adobe Systems, Digital Island, Endemol, Intertrust, Microsoft, and the Motion Picture Association. The chapter provides in-depth longitudinal data on these seven players to show how companies implement DRM processes. After giving a definition of Intellectual Property and Digital Rights Management (first section), the chapter provides a description of the typical Digital Media Management Value Chain Activities and players involved along the different phases examined (second section). An in-depth description of Digital Rights Management processes is discussed in the third section. Digital Rights Management processes are considered in the context of business model and they are distinguished into content processes, finance processes and Rights Management processes. It concludes with a discussion of the model and main benefits generated by the integration of digital rights management (fourth section).*

# Introduction

The burgeoning market for information and entertainment over TV, PC and mobile devices is forcing media operators and content providers to develop their businesses in order to remain competitive. With the availability of more sophisticated content and the increasingly popular trend of peer-to-peer distribution, the requirement for Digital Rights Management (DRM) is becoming essential and the early movers in the operator community are aware of the opportunities they will miss if their DRM solutions are not in place.

Digital Rights Management (Duhl, 2001) poses one of the greatest challenges for multimedia content providers and interactive media companies in the digital age in order to make their interactive products and service catalogues profitable and to face information security management issues.

The importance of protecting digital contents is crucial for content and media rights holders looking to distribute and re distribute their digital contents over more and more digital channels (TV, radio, Internet).

Traditional management of intellectual property rights in digital environments is based on prohibiting access to the content if customer has not presented the proper considerations. This is facilitated by encryption and security measures, and forces the content providers to select business models according to the available technology. Since success in electronic commerce seems to depend on the companies' business models, it is conceded that the equilibrium between technology and the way of doing business should be vice versa (Rosenblatt et al., 2002).

Currently, associated under the term Digital Rights Management (DRM), the domain has developed from an immature consideration of digital products' protection to a new definition of Digital Rights Management, adopted in this study, which covers the identification, description, trading, protection, monitoring and tracking of rights permissions, constraints, and requirements over assorted assets, either tangible or intangible by limiting content distribution (Iannella, 2001).

Moreover, the assignment of these requirements is challenging, as the domain lacks sufficient framework that has a level of abstraction applicable in multiple situations and that describes the definitive characteristics of the domain elements.

The purpose of this study is to analyze the impact generated by the adoption of Digital Rights Management (DRM) processes on the typical digital media management value chain activities trying to analyze the processes in the context of the business model.

After giving a definition of intellectual property and digital rights management (first section), the chapter provides a description of the typical digital media management value chain activities and players involved along the different phases examined (second section). An in-depth description of digital rights management processes is discussed in the third section. The chapter concludes with a discussion of the model and main benefits generated by the integration of digital rights management (fourth section).

# Research Methodology

Given the early stage of the theory development in the field of DRM, the study follows the logic of grounded theory (Glaser & Strauss, 1967) by building the research on a multiple-case study methodology (Eisenhardt, 1989; Stake, 1995; Yin, 1993, 1994). The case study strategy consists of defining the study focus, framework construction, interviews, data collection, and case analysis.

Case studies are frequently utilized to gain a greater depth of insight into organizations and their decision-making processes than is available with large sample surveys (Yin, 1993, 1994). Case studies often have very small sample sizes (Yin, 1993, 1994).

The companies selected are successful players that have adopted DRM processes. These companies are Adobe Systems, Digital Island, Endemol, Intertrust, Microsoft, and the Motion Picture Association. For each company, the study is designed to interview the chief information officer or equivalent executive, and one or two managers in charge.

Interviews were based on a standard set of interview questions. At the time of the interview, additional supporting documents were gathered or requested. Examples include organization charts, annual reports, product reports, planning reports, and Websites containing company and product information.

The chapter provides in-depth longitudinal data on these six players to show how they implement DRM processes.

After giving a definition of intellectual property and digital rights management (first section), the chapter provides a description of the typical digital media management value chain activities and players involved along the different phases examined (second section).

Digital rights management processes are considered in the context of a business model and they are distinguished into content processes, finance processes and rights management processes (third section). The chapter concludes with a discussion of the model and main benefits generated by the integration of digital rights management and propose the most interesting directions for future research (fourth section).

*Table 1. Sample of six companies*

| Case No: | Case 1 | Case 2 | Case 3 | Case 4 | Case 5 | Case 6 |
|----------|--------|--------|--------|--------|--------|--------|
|          | Endemol | Digital Island | Adobe Systems | Intertrust | Motion Picture Association | Microsoft |
| Industry: | Multimedia Content Provider | Internet Service Provider | Internet Service Provider | Technology and platform provider | Association | Software Provider |

# Intellectual Property:
## Definition

Intellectual property refers to all moral and property rights on intellectual works. Intellectual property rights (IPRs) are bestowed on owners of ideas, inventions, and creative expressions that have the status of property (Scalfi, 1986). Just like tangible property, IPRs give owners the right to exclude others from access to or use of their property.

Article 1, Section 8 of the *U.S. Constitution* emphasises the utilitarian nature of intellectual property: "To promote the progress of science and useful arts, by securing for limited times to authors and inventors the exclusive right to their respective writings and discoveries".

The first international treaties covering intellectual property rights were created in the 1880s and they are administered by the World Intellectual Property Organization (WIPO), established in 1967. The new legislative frameworks, in response to 1996 WIPO treaties, include DMCA in the U.S. and European Copyright and eCommerce Directives, along with new legislation such as special treatment for "technological protection methods," "pragmatic exception for «transient copies», notice and take down" procedures for alleged breaches of copyright.

The newly revealed physics of information transfer on the Internet has changed the economics and ultimately the laws governing the creation and dissemination of intellectual property.

Internet poses challenges for owners, creators, sellers and users of intellectual property, as it allows for essentially cost-less copying of content. The development of Internet dramatically changes the economics of content, and content providers operate in an increasingly competitive marketplace where barriers to publishing are disappearing and much content is distributed free (see the Napster phenomenon). Concepts of territoriality are meaningless on the network, even if they are very significant for business in the physical world.

There are many issues that organizations need to address to fully realize the potential in their intellectual property. They can be summarized in the following:

- *Ownership:* clear definition of who owns the specific rights and under what circumstances;

- *Distribution:* definition of the distribution strategy (small trusted group or the mass market);

- *Protection:* definition of the content the organizations need to protect and the level of protection required;

- *Globalization:* because of the slow harmonization across countries regarding protection of intellectual property rights, what is acceptable in one country may have legal implications in another;

- *Standards:* understanding what standards are in development and how these may affect system development.

# Digital Rights Management:
## Functional Architecture

From a technologist's perspective, digital rights management refers to the set of techniques for specifying rights and conditions associated with the use and protection of digital content and services (Damiani, 2003).

From the business perspective adopted in this study, the term *digital rights management* covers the description, identification, trading, protection, monitoring and tracking of all forms of rights uses over both tangible and intangible assets, including management of rights holders relationships (Iannella, 2001).

At the heart of any DRM technology is the notion of a rights model.

Rights models are schemes for specifying rights to a piece of content that a user can obtain in return for some consideration, such as registering, payment, or allowing usage to be tracked.

*Digital rights management* can be defined as the secure exchange of intellectual property, such as copyright-protected music, video, or text, in digital form over channels such as the Web, digital television, digital radio, the much talked 3G (third-generation) mobile or other electronic media, such as CDs and removable disks.

The technology protects content against unauthorized access, monitors the use of content, or enforces restrictions on what users can do with content.

Digital rights management allows organizations that own or distribute content to manage the rights to their valuable intellectual property and package it securely as protected products for digital distribution to a potentially paying, global audience.

DRM technologies provide the basic infrastructure necessary for protecting and managing digital media, enterprise-trusted computing, and next generation distributed computing platforms, and they allow content owners to distribute digital products quickly, safely, and securely to authorized recipients.

The digital media management value chain can be described by the following main areas, which play a key role in building digital rights-enabled systems:

1.  *Intellectual Property (IP) Asset Creation and Acquisition:* This area manages the creation and acquisition of content so it can be easily traded. This includes asserting rights when content is first created (or reused and extended with appropriate rights to do so) by various content creators/providers. This area supports:

    •   **rights validation:** to ensure that content being created from existing content includes the rights to do so;

    •   **rights creation:** to allow rights to be assigned to new content, such as specifying the rights owners and allowable usage permissions;

    •   **rights workflow:** to allow for content to be processed through a series of workflow steps for review and/or approval of rights (and content).

2.  *Intellectual Property Media Asset Management:* After the finished content is bought, this area manages and enables the trade of content. The digitalization of the television signal and the storage of the materials that have been purchased need to manage the descriptive metadata and rights metadata (e.g., parties, uses, payments, etc.). This area supports:

    •   **repository functions:** to enable the access/retrieval of content in potentially distributed databases and the access/retrieval of metadata;

    •   **trading functions:** to enable the assignment of licenses to parties who have traded agreements for rights over content, including payments from licensees to rights holders (e.g., royalty payments).

3.  *Intellectual Property Asset Delivery Management:* This area manages the distribution and usage of content through different platforms (TV, radio, Web, 3G mobile) once it has been traded. This includes supporting constraints over traded content in specific desktop systems/software. This area supports:

- **permissions management:** to enable the usage environment to honor the rights associated with the content (e.g., if the user only has the right to view the document, then printing will not be allowed);
- **tracking management:** to enable the monitoring of the usage of content where such tracking is part of the agreed to license conditions (e.g., the user has a license to play a video 10 times).

# Digital Rights Management:
## Value Chain Activities

After defining the functional architecture, DRM value chain is described, at a first level, identifying the activities supported by each segment and the players involved. At a second level digital rights management processes are analyzed in the context of the digital content management value chain and the impact of these processes on the business model.

The following findings refer to an in-depth longitudinal survey on six players (Adobe Systems, Digital Island, Endemol, Intertrust, Microsoft and the Motion Picture Association) trying to understand how they implement DRM processes.

The digital rights management value chain can be described by six main segments (see Figure 1): contract & rights management, rights information storage, licence management, persistent content protection, clearing house services, billing services (Burke, 2002).

Each segment along the value chain is characterized by specific activities:

1. *Contract and Rights Management:* the registration of contract terms and rights, tracking of usage, and payment of royalties (residuals);

2. *Rights Information Storage:* the storage of rights information (e.g., play track five times) and usage rules (e.g., if they have a UK domain and have paid) as well as rights segmentation and pricing structures;

3. *License Management:* the management and issuing of licenses in line with the rights and conditions. Without a license the consumer cannot use the content;

4.  *Persistent Content Protection:* the use of encryption, keys, and digital watermarking to securely package digital content;

5.  *Clearing House Services:* managing and tracking the distribution of the packaged content and the license in line with the defined rights information;

6.  *Billing Services:* charging consumers for purchased content and payment to parties within the value chain.

Various players in the market are affected by digital rights management, in particular all those operators involved with ideation, creation, realization, enabling and handling of contents (i.e., Endemol). The role of these operators is to provide contents successively broadcast on the networks channels, radio, the Internet and other media. The different players involved along the value chain can be summarized in the following:

- *Content Author/Creator* (e.g., Canal+, Reuters): or any other media or non-media company that produces content for internal or external use;

- *Content Publisher/Aggregator* (e.g., IPC Magazines, Flextech Television, BskyB): buys various content and aggregates it into channels aimed at a particular lifestyle or niche;

- *Content Distributor* (e.g. W.H. Smith News, BskyB, NordeaTV): sends the content from the service head-end to the user device and back and asserts that streams are delivered on time;

- *Service Provider/eTailer* (e.g., T-Online.com, Yahoo.com): provides various individual elements of customer service, billing function or aspects of the physical network.

In the first phase of the DRM value chain *(Contract & Rights Management)*, after the author has created the content, the aggregator packages it in a container that provides persistent protection, enforcing the rights that the author has granted. This may be written as an applet that travels with the content that will be encrypted.

In the second phase *(Rights Information Storage),* the aggregator specifies the rights that apply to the content using products such as IBM, EMMS or Microsoft, which encode the rights using an XML-based standard such as XrML or XMCL.

In the third phase *(License Management)*, the consumer purchases the rights to use the content, the eTailer obtains the content from the distributor and requests a license from the content clearing house. The license may well be written in XML and may travel with the packaged content or separately.

The fourth phase is *Digital Asset Protection.* The consumer cannot access the content without the license. The Media Player, for example, Real Video Player, interprets the license and enforces the rights granted to the consumer. That may include how many times or for how long they can access the asset, and whether they can duplicate it or "rip" CDs from it.

With reference to the digital content management value chain described above, DRM processes play an important role in the encode activity (Contract Management & Rights Storage), management and archive assets (Encrypt/ Package Content Asset Protection), management workflow (Key & License Generation) and exploit (Key/License Management & Clearing-house Services) (Figure 1).

At a second level of analysis, digital rights management processes described above can be understood in the context of the business model.

*Figure 1.  DCM value chain activities & DRM processes*

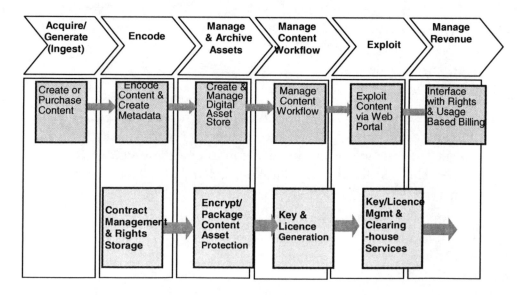

*Figure 2. The processes in the context of the business model*

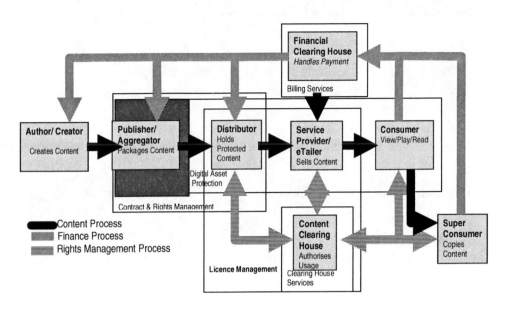

All the processes analyzed can be grouped in three broad categories (see Figure 2):

1.  **content processes:** include all systems and processes that enable the creation/capture, formatting, packaging, protection and storage of trusted channel content, control of when and how content is released, where and when it gets used and how long it remains there;

2.  **finance processes:** include all systems and processes (payments) between the financial clearing house and the other players along the value chain (author/creator, publisher, distributor, consumer);

3.  **rights management processes:** include all content clearing house services which authorize the usage of rights to distributors, service providers, and consumers. They include all processes between the content clearing house and the other players along the value chain (distributor, service provider e-tailer, consumer).

# Digital Rights Management Benefits

This section describes the benefits generated by digital rights management.

As described in the first paragraph the methodology for this research is case study (Stake, 1995; Yin, 1993, 1994). The companies interviewed are successful players that have adopted DRM processes. These companies are Adobe Systems, Digital Island, Endemol, Intertrust, Microsoft, and the Motion Picture Association. The following benefits emerge from the interviews conducted.

The first main area of analysis concerns all benefits related to Contract & Rights Management.

Digital rights management allows a better management of bought-in content rights to maximize return on investment, and it avoids potentially expensive misuse.

A better management of content is allowed also by means of the creation of what-if scenarios for potential new revenue streams on the basis of the cost and ownership of content rights. New contents can no longer be constrained to budget on a cost basis, but they can now budget on the basis of forecast rights revenue.

Digital rights management generates benefits also with reference to the sale of rights, allowing a retained control of the sale of content rights and the conditions of sale in mass or niche distribution environments.

DRM allows the establishment of flexible business models for digital content sales (e.g., rental or purchase, play or edit, burn to CD, or just play online).

Segmentation of content allows the creation of different versions with different rights and conditions for different markets.

Information that has previously been stored in a vast number of separate databases can now be merged, sorted, and analyzed, resulting in the creation of a personal profile or data image of a subject based on his or her electronic data composite (Tapscott, 1999).

Consumer acceptance will determine the success of the market. The critical factor for vendors is to address the right market (publishing, audio, video and software).

As declared by some players (Table 2-3), it allows persistent content protection, the opportunity to distribute the content in different markets, and it allows usage monitoring.

New business models need to be implemented.

## Table 2. Main economic benefits of DRM

| Case 3 | Case 4 | Case 5 |
|---|---|---|
| **Endemol** | **Intertrust** | **Microsoft** |
| DRM is key for **financial success,** but lots need to be done:<br><br>- Security is a big issue but does not yield profits!<br><br>- Current market models "fail" in the interactive world where the customer is central.<br><br>- An increasingly demanding customer and the vast amount of exploitation options will challenge parties even further. | **- Persistent Protection**<br>No matter how much your content is copied you always get paid.<br>**- Superdistribution**<br>You can encourage people to pass content around. In effect they can become distributors and our software allows them to be financially rewarded for this.<br>**- Usage Monitoring**<br>All content usage is monitored – you can see how any types of people use your content (with permission).<br>**- Micro-Payments**<br>Our system practically supports small transactions. | **New revenue opportunities**<br>- Direct response content linked to a broadcast stream<br>- Enhanced TV with e-commerce<br>- Targeted advertising and permission-based direct mail<br>**New Markets**<br>- Internet Services<br>- Education, Distance Learning<br>- Home Data Services<br>- Business Services<br>- Private Data Services |

## Table 3. Technology issues

| Case 1 | Case 4 |
|---|---|
| **Endemol** | **Intertrust** |
| - Technical possibilities for large audiences on interactive media are still to be examined closely!<br><br>- DRM is needed, but which system solves all our problems?<br>- Investment?<br>- Future needs?<br>- Working billing solutions across various platforms.<br><br>- Technique & new processes including DRM need to be used, but the end consumerdoes not need to be bothered. | - Conventional computing not designed to protect rights in information. Trusted transactions take place on servers. Access to the servers is protected and the network is protected, but not the content itself.<br><br>- Our technology embodies a fundamental addition to computer science. It can protect the content and associate rules that are persistently managed. Content can flow freely.<br>a. We manage from the *user* environment.<br>b. We believe this will create the *next great inflection point*.<br>c. Result is that each device in the system is able to process protected transactions.<br>d. Efficient because using processing power that is out there. Can process any size transaction, whereas Visa can only process over $10.00.<br><br>- Content is always protected and managed for online and offline use. Can permit access to content even when offline and rules are still applied. |

# Conclusion

In this chapter I discussed how digital rights management allows organizations to manage the rights to their valuable intellectual property and package it securely as protected products for digital distribution.

It is not solely about technology; digital rights management works across the people, processes and technology boundaries.

The key issues emerging from the analysis include:

- handling of complex sets of rights within each asset;
- rights licensing and management, and digital rights protection;
- understanding and design of revenue generation and collection models;
- standards - flexible rights languages and content formats;
- globalization - territorial issues, both legal and commercial;
- ownership of rights.

Digital rights management is emerging as a formidable new challenge, and it is essential for DRM systems to provide interoperable services. Industry and users are now demanding that standards be developed to allow interoperability so as not to force content owners and managers to encode their works in proprietary formats or systems.

The market, technology, and standards are still maturing. Digital rights management should be considered an integral part of a company's digital media management framework.

# References

Alattar, A.M. (2000, January). 'Smart Images' Using Digimarc's watermarking technology. *Proceedings of the IS&T/SPIE's 12th International Symposium on Electronic Imaging,* San Jose, CA, *3971*(25).

Anderson, L.C., & Lotspiech, J.B. (1995). Rights management and security in the electronic library. *Bulletin of the American Society for Information Science, 22*(1), 21-23.

Association of American Publishers Copyright Committee. (2000). *Contractual licensing, technological measures and copyright law*. Washington, D.C.: Association of American Publishers.

Association of American Publishers Rights and Permissions Advisory Committee. (2000). *The new & updated copyright primer: A survival guide to copyright and the permissions process*.

Burke, D.C. (2002). *Digital rights management: From zero to hero?* Cap Gemini Ernst & Young, speech at IBC Nordic Euroforum Conference, Stockholm, Sweden.

Burns, C. (1995). *Copyright management and the NII: Report to the Enabling Technologies Committee of the Association of American Publishers*. Washington, D.C.: Association of American Publishers.

Chaudhuri, A.K. et al. (1995, May-June). Copyright protection for electronic publishing over computer networks. *IEEE Network, 9*(3), 12–20.

ContentGuard, I. (2001). *eXtensible rights Markup Language (XrML) 2.0 Specification* [Online]. Available: *http://www.xrml.com, [26.8.2002]*

Duhl, J. (2001). *Digital Rights Management (DRM): A definition*. IDC.

Duhl, J. (2001). *DRM Landscape: Technologies, vendors, and markets*. IDC.

Gervais, D.J. (1997). Electronic rights management and digital identifier systems. *Journal of Electronic Publishing, 4*(3). Ann Arbor, MI: University of Michigan Press.

Iannella, R. (2001). Digital Rights Management (DRM) architectures. *D-Lib Magazine, 7*(6).

Iannella, R. (2002). *Open Digital Rights Language (ODRL) Version 1.1. IPR Systems Pty Ltd*. Available: http://www.odrl.net/1.1/ODRL-11.pdf, *[31.8.2002]*

Interactive Multimedia Association. (1994). *Proceedings: Technological strategies for protecting intellectual property in the networked multimedia environment*. Annapolis, MD: Interactive Multimedia Association.

Kahin, B., & Arms, K. (Eds.). (1996). *Forum on technology-based intellectual property management: Electronic commerce for content*. Special issue of *Interactive Multimedia News, 2*.

Luoma, E., Tiainen S., & Tyrväinen, P. (2003). Integrated domain model for Digital Rights Management. In M. Khosrow-Pour (Ed.), *Information technology and organization trends, issues, challenges and solutions.* Hershey, PA: Idea Group Inc.

Lyon, G. (2001). *The Internet marketplace and Digital Rights Management.* National Institute for Standards and Technology.

Miron, M., De Martini, T., Wang, X., & Gandee, B. (2001, December). *The language for digital rights.* Presented at MPEG 58[Th] Meeting, Pataya, Thailand.

Plassard, M.-F. (1998). Functional requirements for bibliographic records. Final Report, IFLA Study Group on the Functional Requirements for Bibliographic Records. Available: http://www.ifla.org/VII/s13/frbr/frbr.pdf

Risher, C., & Rosenblatt, B. (1998) The digital object identifier - An electronic publishing tool for the entire information community. *Serials Review, 24*(3/4), 13-21. Stamford, CT: JAI Press, Inc.

Rosenblatt, B. (1996). *Two sides of the coin: Publishers' requirements for digital intellectual property management.* Inter-Industry Forum on Technology-Based Intellectual Property Management, Washington, DC.

Rosenblatt, B. (1997). The digital object identifier: Solving the dilemma of copyright protection online. *Journal of Electronic Publishing, 3*(2). Ann Arbor, MI: University of Michigan Press.

Rosenblatt, B., Trippe, B., & Mooney, S. (2002). *Digital rights management: Business and technology.* New York: M&T Books.

Rust, G., & Bide, R. (2000). *<indecs> metadata framework: Principles, model and dictionary.* Indecs Framework Ltd. Available: http://www.indecs.org/pdf/framework.pdf

Scalfi, G. (1986). *Manuale di Diritto Privato, UTET,* Torino, 61.

Silbert, O. et al. (1995). DigiBox: A self-protecting container for information commerce. *Proceedings of the First USENIX Workshop on Electronic Commerce,* New York, NY.

Souzis, A. et al. (2000). *ICE implementation cookbook: Getting started with Web syndication.*

Stefik, M. (1996). Letting loose the light: Igniting commerce in electronic publication. *Internet dreams: Archetypes, myths, and metaphor.* Cambridge, MA: MIT Press.

Stefik, M. (1999). The Internet edge: Social, technical, and legal challenges for a networked world. Cambridge, MA: MIT Press.

Tapscott, D. (1999). Privacy in the digital economy. *The digital economy* (p. 75). McGraw Hill.

Vaidhyanathan, S. (2001). Copyrights and copywrongs: The rise of intellectual property and how it threatens creativity. New York: NYU Press.

Van Tassel, J. (2001) *Digital content management: Creating and distributing media assets by broadcasters.* Washington, D.C.: NAB Research and Planning Department. Available from National Association of Broadcasters at (202) 429-5373.

Vonder Haar, S. (2001). *Digital rights management- Securing new content revenue streams.* Yankee Group Report.

# About the Authors

**Marian Quigley** (HDTS, Art & Craft, Melbourne State College; B.A., Chisholm Inst.; Ph.D., Monash University) is a senior lecturer and director of Research and Postgraduate Studies in the School of Multimedia Systems, Faculty of Information Technology, Monash University, Australia. Marian has published several articles and presented a number of papers relating to social and ethical issues in information technology, particularly in relation to youth. She is currently completing a book on the effects of computer technology on Australian animators.

*  *  *  *  *

**Rasool Azari** is associate professor in the School of Business at the University of Redlands (USA) and previous associate chair of the Department of Management and Business. He holds a Doctor of Science in Engineering Management and MBA in International Business from George Washington University, and a Master of Science in Electrical Engineering from UCLA. Since 1996 he has been a consultant to the United States Department of Commerce/National Institute of Standards and Technology (NIST). He has numerous publications on the impact of socioeconomic factors on technological change. His research interests include supply chain management, the influence of technological change, organizational transformation, and technology management and policy.

**Vaughn R. Christie** is a recent Master's of Science graduate from the Department of Telecommunications & Networking Technology in the School of Computer Technology at Purdue University, USA. As a graduate student,

Vaughn served as teaching assistant for Wide Area Networking and Network Management courses and was responsible for developing, maintaining and building lab-based instruction modules for many of the school's undergraduate students. In addition to corporate information security concerns, Vaughn's research interests include system administration, network design, and finding the delicate balance between industry need and the business value of information technology and strategic information systems.

**Jack S. Cook** is a professor, speaker, author and consultant. He is an associate professor of Management Information Systems at the Rochester Institute of Technology (RIT) (USA). His areas of expertise include e-commerce, information systems and production/operations management. Dr. Cook's extensive experience teaching and training over the last two decades includes more than 90 conference presentations and numerous journal articles. He has an entertaining and engaging approach and is known for bringing theories to life. Dr. Cook is a certified fellow in Production and Inventory Management (CFPIM). His education includes a Ph.D. in Business Administration, an M.S. in Computer Science, an MBA, an M.A. in Mathematics, and a B.S. in Computer Science. To learn more, visit: www.sizzlingsolutions.com.

**Laura Cook** works as a technology support professional for the Computing & Information Technology Department at the State University of New York at Geneseo (USA). She manages all technology support for seven departments. She has a Bachelor's in Business Technology from the Rochester Institute of Technology (RIT). Laura is Webmaster and on the board of directors for the Rochester, New York chapter of APICS.

**Neil F. Doherty** is a senior lecturer in Information Systems in the Business School at Loughborough University, UK. In addition to information security, his research interests include the interaction between organisational issues and technical factors in information systems development, understanding the reasons for failures of information systems projects, strategic information systems planning and e-commerce. Neil has had papers published, or forthcoming, in a range of academic journals, including: *European Journal of Information Systems, Journal of Information Technology, Journal of Strategic Information Systems, IEEE Transactions in Engineering Management, Journal of Business Research, Journal of End User Computing* and *Information & Management*.

**Alfreda Dudley-Sponaugle** is a lecturer in the Department of Computer and Information Sciences at Towson University (USA). She currently teaches the computer ethics courses in the curriculum. Professor Dudley-Sponaugle presented and published at the following conferences: Proceedings of the Eighth Annual Consortium for Computing Sciences in Colleges – Northeastern Conference; IEEE 2002 International Symposium on Technology and Society (Social Implications of Information and Communication Technology); and the Forty-first ACM Southeast Regional Conference. She is co-advisor to the National Black Science and Engineering chapter at Towson. She is serving on the Business and Education Steering Committee on a proposed NSF grant entitled "The Grace Hopper Scholars Program in Mathematics and Computer Science". Professor Dudley-Sponaugle has participated as a faculty representative/participant at The Howard County Public School System: Mathematics, Science and Technology Fair.

**Craig Fisher** is an associate professor of Information Systems at Marist College in Poughkeepsie, New York, USA. He received his Ph.D. in Information Science in 1999 from the University at Albany (SUNY). He enjoyed a career at IBM, where he advanced through various information systems middle management positions. He now teaches problem solving and programming, information systems policy, data quality and systems design. His papers have appeared in *Information & Management* and *Information Systems Research* journals as well as various conferences.

**Heather Fulford** is a lecturer in Information Systems in the Business School at Loughborough University (UK). Her research interests include security management in large and small enterprises, electronic commerce adoption, Website design, and knowledge management. She is currently managing an EPSRC-funded project investigating the adoption of IT by UK SMEs, and has also gained government funding for an e-commerce adoption project.

**James E. Goldman** is professor and associate department head in the Department of Computer Technology at Purdue University (USA), where he founded the Telecommunications and Networking Technology program. Jim has more than 20 years of experience in telecommunications strategy, network engineering, and project management of regional, national, and global networking projects. Jim is a CISSP (certified information systems security professional) with advanced training in computer forensics. He is an internationally published author with market leading textbooks in data communications and networking. Among Jim's areas of research and publication are municipal telecommunications strategy, information technology economics and investment strategy, and network security investment strategy.

**Andrzej T. Jarmoszko** is an associate professor in the Department of Management Information Systeams, School of Business, Central Connecticut State University (USA). His primary teaching areas are systems analysis and design and data communications and networking. Prior to joining CCSU faculty, Dr. Jarmoszko was Manager for Strategic Planning at a major mobile communications company in central Europe. His research interests include information systems curriculum, mobile information systems, aligning knowledge management with the strategy process, and strategic management in the communications industry.

**Laura Lally** is an associate professor at the Frank G. Zarb School of Business at Hofstra University (USA). She holds a Ph.D. in Information Systems from the Stern School of Business at New York University, and an M.B.A. from Baruch College of the City University of New York. She has published articles in *Decision Sciences,* the *Journal of Global Information Management,* the *Journal of End-User Computing,* the *Information Society,* and the *Journal of Business Ethics.* She has received two National Science Foundation Grants to support her research in applying normal accident theory and the theory of high reliability organizations to information technology.

**Jonathan Lazar** is an assistant professor in the Department of Computer and Information Sciences, and an affiliate Professor in the Center for Applied Information Technology at Towson University (USA). He is the author of the book *User-Centered Web Development* and editor of the book *Managing IT/Community Partnerships in the 21ˢᵗ Century.* He is on the editorial board of the *Information Resource Management Journal* and is Associate Editor of the *Journal of Informatics Education and Research.* Dr. Lazar regularly presents and publishes papers on the topic of Web usability and accessibility. He has served on the program committee for conferences sponsored by the ACM Special Interest Group on Computer-Human Interaction (CHI) and for the Information Resource Management Association. He is the 2002 winner of the "Excellence in Teaching" award in the College of Science and Mathematics at Towson University.

**Leslie Leong** is associate professor of M.I.S. at Central Connecticut State University (USA). Her teaching interest includes e-commerce, database management systems, and Web development technologies. She has taught both the undergraduate and graduate level in M.I.S. Her major research interests include I.S. security, system usage, e-commerce, and database systems. She has published in numerous regional and national proceedings and academic journals. Dr. Leong is a member of the Association of Information Systems, Decision Sciences Institute, and Information Resources Management Association.

**Margherita Pagani** is head researcher for New Media&TV-lab inside I-LAB Centre for Research on the Digital Economy of Bocconi University (Milan - Italy). She is adjunct professor in Management at Bocconi University in Milan. She has been visiting scholar at Sloan – MIT. She has written three books on digital interactive television and wireless and mobile networks and many publications and reports about interactive television, digital convergence, content management, digital rights management discussed in many academic conferences in Europe and USA. Dr. Pagani serves on the editorial advisory board for *Journal of Information Science and Technology* – USA. She has worked with RAI Radiotelevisione Italiana and as Associated Member of the Permanent Forum of Communications (work group "Digital Terrestrial") for Ministry of Communications (Italy). E-mail: margherita.pagani@uni-bocconi.it

**Helen Partridge** is a lecturer in the School of Information Systems at the Queensland University of Technology (QUT) (Australia). Helen is involved in teaching both undergraduate and postgraduate students in the areas of information management, professional practice, reference and information service, information resources and information literacy. Before joining QUT, Helen spent several years working in the public library sector. Helen in currently completing a Ph.D. through the Centre for Information Technology Innovation (CITI) at QUT. The Ph.D. explores the psychological perspective of the digital divide within community.

**Malcolm R. Pattinson** was an IT practitioner and consultant for 20 years prior to joining the University of South Australia in 1990. He teaches IS security management and related topics and supervised post-graduate research students. He has conducted many IT/IS consultancy and research projects and is currently working with the South Australian government in reviewing their IS security controls. His research interests are in the areas of IS security evaluation and metrics and he has published and presented numerous papers on this topic. He has been a member of ISACA for eight years and a CISA for six years and is currently the Australian representative for ISACA's international academic liaison committee. He is also a member of IFIP TC11 Working Group 11.2 – Small Systems Security.

**James Pick** is professor in School of Business at the University of Redlands (USA), chair of its faculty assembly, and former department chair of Management and Business. He holds a B.A. from Northwestern University and Ph.D. from University of California, Irvine. He is the author of seven books and 105 scientific papers and book chapters in the research areas of management information systems, geographic information systems, environmental systems, population, and urban studies. He has received several university and external research and teaching awards, and in 2001 was senior Fulbright scholar in Mexico.

**Geoffrey A. Sandy** is head, School of Information Systems of the Faculty of Business and Law, Victoria University, Melbourne, Australia. He holds a Doctor of Philosophy from RMIT University and a Bachelor's and Master's in Economics from Melbourne University. Geoff is a member of the Australian Computer Society, the Australian Institute of Computer Ethics and Electronic Frontiers Australia. He has more than 27 years experience in university teaching, research and consultancy and six years in private sector experience, where he worked for International Harvester and the ANZ Bank. His major research interests are Internet censorship, filter software, computer use policies, requirements engineering and adding value via the Internet. He has published widely in these and other areas.

**Barbara A. Schuldt** is an associate professor in the Management Department at Southeastern Louisiana University (USA). Her research focus has been on ethical use of information systems and technology, data capture via technology and the teaching management information systems. She is active in the Information Resource Management Association and reviews for *Information Resource Management Journal, Journal of Database Management*, and *Decision Sciences Institute.*

**Trisha Woolley** is currently pursuing her master's degree in Information Systems and Business Administration at Marist College in Poughkeepsie, New York, USA. She has performed extensive tutoring in mathematics and is currently an adjunct lecturer at Dutchess Community College. Ms. Woolley has performed data quality analysis for the Dutchess County Office for the Aging and has presented a paper at the Information Resource Management Association 2003 International Conference.

# Index

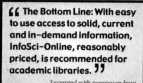

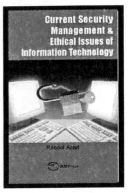

# Social, Ethical and Policy Implications of Information Technology

Linda L. Brennan, PhD, Mercer University, USA
Victoria Johnson, DPA, Mercer University, USA

Legal and ethical issues have become a standard part of engineering and business schools' curricula. This has not been the case for computer science or management information systems programs, although there has been increasing emphasis on the social skills of these students. This leaves a frightening void in their professional development. Information systems pose unique social challenges, especially for technical professionals who have been taught to think in terms of logic, structures and flows. **Social, Ethical and Policy Implications of Information Technology** focuses on the human impact of information systems, including ethical challenges, social implications, legal issues, and unintended costs and consequences.

*ISBN 1-59140-168-2 (h/c)* • US$74.95 • *ISBN 1-59140-288-3 (s/c)* • US$59.95
• *320 pages* • *Copyright © 2004*

"By creating awareness and offering analytical frameworks for a variety of issues stemming from information systems' implementations, this book can contribute to the development of the technical professionals' ability to cope — and perhaps avert —problems."

- *Linda L. Brennan, Mercer University, USA*
*Victoria Johnson, Mercer University, USA*

**It's Easy to Order! Order online at www.idea-group.com or
call 717/533-8845 x10**
Mon-Fri 8:30 am-5:00 pm (est) or fax 24 hours a day 717/533-8661

 **Information Science Publishing**
Hershey • London • Melbourne • Singapore

*An excellent addition to your library*